THE KILT BEHIND THE CURTAIN

A SCOTSMAN IN CEAUȘESCU'S ROMANIA

RONALD MACKAY

WEE DRAM PUBLICATIONS

Copyright © 2020 by Ronald Mackay

ISBN: 9798683193768

Published by Wee Dram Publications

Formatted by AntPress.org

All rights reserved.

No part of this book may be reproduced in any form or by any electronic or mechanical means, including information storage and retrieval systems, without written permission from the author, except for the use of brief quotations in a book review.

This memoir is dedicated to:

'M' who dared and won

Pearl Mackay, my ever-supportive mother

My wife, Viviana Carmen Galleno Zolfi for suggesting I write

Dino Sandulescu, for sharing and enduring.

CONTENTS

1. To Bucharest aboard the Orient Express	7
2. Early Days	10
3. How All This Began	15
4. Travel Warning	20
5. The Prahova Valley	23
6. Getting down to Work	29
7. Who's who?	39
8. Britain's Bizarre Book Presentation	49
9. Sour Milk?	54
10. All the Better to Know You	57
11. Daily Shopping	61
12. The British Commissary	65
13. Taking stock	69
14. A Long Wheelbase Land Rover	74
15. Pearl's Arrival	82
16. Transylvania and Moldavia	88
17. Painted Monasteries	95
18. Home to Bucharest	100
19. Year One Draws to an End	106
20. Summer of '68	120
21. Security Matters	125
22. The Post-Prague Spring Year in Romania Begins	133
23. American Counterparts	138
24. Regarding Romance	141
25. The Lipoveni and Murfatlar	146
26. Founding Friendships	149
27. And More	160
28. Romance	166
29. Why walk?	174
30. 'M' and 'D' Combined	180
31. Skiing at Sinaia	185
32. Scholarships to Britain	190
33. Mutton Chops with Petru	197
34. Scholarship Candidates	200

35. Alexandru from Alexandria	203
36. Cross-Border Travel	206
37. A Troop of Tanks	218
38. More Travels with Pearl	230
39. Danube Delta	236
40. Siebenbürgen	245
41. Ivan Deneş	256
42. Sitting Ducks	268
43. Tiberiu Stoian and the School for Spies	275
44. What Plans, Ronald?	278
45. Formal Goodbyes	283
46. Summer of '69	286
47. Romanian Fallout	289
Acknowledgments	291
About the Author	293

1

TO BUCHAREST ABOARD THE ORIENT EXPRESS

Setting Out

Any train journey is exciting; even more so if your trip takes two nights and three days. But if it is the 1960s and you're travelling on a challenging mission into a communist country deep behind the Iron Curtain on the infamous Orient Express, the experience is intoxicating.

Arrival at Gara de Nord, Bucharest

In the summer of 1967, I was met at Bucharest by an official from the British Embassy. She knew that after three days on the Orient Express, all I wanted was to be taken to the apartment assigned to me by the Romanian Government and left to myself. With few words, she drove me through wide, traffic-free boulevards, past elegant and distinctive architecture until we reached an anonymous apartment building.

"Please visit the Cultural Attaché at the British Embassy on Monday at two." She handed me a set of keys, a roll of Romanian banknotes and drove off. I was grateful to be alone.

That evening, I was impatient to explore. On the steps of my apartment building, I quietly took stock. The block stood on the corner of two wide, treelined boulevards entirely devoid of cars. Passengers hovered, waiting for the occasional tramcar or trolleybus. There was no turn-taking. People lurched to the doors fighting to board before others alighted. They thrust with arms and knees and legs and hands, showing no deference to age or gender.

The display windows of a line of shops, except one, were empty of goods. "Gospodina" offered a display of appetizing cooked foods in trays. I'd no desire to spend my first night in Bucharest eating alone in my apartment. What I wanted was to see Romanians, how they dressed, how they talked and laughed, so I went exploring. Since the new streets were laid out on a grid, I knew I could find my way back.

My nose leads me to a busy neighbourhood restaurant. At the door, I pause to take in the comings and goings. I enter, sit at one of the few empty tables, and consult my Romanian phrasebook. Noisy clients eat and drink. Waiters bustle ignoring me. Tall bottles of beer, plates of brown bread, pickles, garlic cloves and cheese, the smell of warm evening cheer.

To a passing waiter, I call out the words I'd practised, "O bere, vă rog!" "Give me a beer, please." Without pausing, he utters something I hear as "Bere nu mai există!" I guess at the translation as, *"Beer doesn't exist"*.

Beer doesn't exist? These bottles had contained beer. Ah! Romanian humour, I decide. To inject levity into his unending work, the waiter denies the very existence of beer in his bar. Expecting a bottle and a glass to be delivered with a wink, I wait. Five minutes pass. Again, I call: "A beer please!" Again comes the response: "Bere nu mai există!"

After my third request, the waiter tramps determinedly to the counter, picks up a tall bottle, brings it back to my table and ostentatiously turns it upside-down.

"Bere numai există, Tovarăşe!" "We're out of beer, Comrade!"

Such was my introduction to the inexplicable shortages that

plagued Romania under communism. Simple basics produced, within the very country, like milk or potatoes or cheese suddenly "ran out" and therefore "no longer existed". Like Romanians I became accustomed.

Nevertheless, in that neighbourhood restaurant, on my first enchanted evening in Bucharest, I enjoyed a plate of flavoursome rye bread, crisp pickles, garlic cloves, and delicious cheese accompanied by a glass of breath-catching țuică, the local plum brandy 40% proof. That glass made finding my way back to my apartment more challenging, but I slept soundly.

Hammer-and-Sickle, adopted by Lenin in 1917 as the Soviet symbol. Widely adopted by the Romanian Communist Party.

2

EARLY DAYS

On my first full day in the city, I walked to Bucharest University's Faculty of Foreign Languages and Literatures on Pitar Moş, where I would be teaching within the month. Fighting to board a trolleybus held no appeal. It was autumn, the trees lining the boulevard were changing colour. Walking was a pleasure.

How city-people dressed reminded me of Scotland in the 1940s. They wore unfashionable, formal, dark clothes, the men in shirt, trousers, jacket and tie. Suits were common. The women, distinctly attractive, wore dresses or skirts and sweaters. Country people dressed like small farmers or peasants, their women in heavy skirts and waistcoats made of sheepskin or woven wool. Both sexes were weather-beaten and most of the men wore handsome, high, karakul caps. The colours and patterns could be breathtaking.

I crossed the Dâmboviţa River. Rivers running through cities always attract. The dark water of the Dâmboviţa flowed slowly within its canalised banks, reminding me of London's canals. Totally different however, from London or anywhere else I had lived, were the striking Romanian Orthodox Churches. Most appeared closed and though intact, poorly maintained.

As I passed one of the more picturesque churches an older woman covertly crossed herself using a single finger on her lips. She was upset that I had caught the gesture. To allay her fears, I crossed myself openly. She hurried on, head down.

The closer I approached to the centre, the more grand became the architecture. Old Bucharest was handsome, elegant. The dearth of traffic and especially of private cars made the walk more pleasant. The few vehicles there were tended to be black, enormous, and driven by serious chauffeurs wearing cloth caps. A very few sleek Mercedes, similarly curtained, protected the identity of privileged passengers, no doubt officials of the Romanian Communist Party. Immaculately uniformed traffic policemen raised white-gloved hands to guarantee the limousines unobstructed passage. At first, I stared, but because nobody else did, I was attracting attention, so I stopped staring, wanting to fit in, to become invisible.

In the heart of the city, the sidewalks were busier. Pedestrians would stare at me curiously but glance away if I caught their eye. My height and general appearance was no different from others so I decided it was my non-Romanian clothes. Through the enormous open casement windows of the Hotel Lido I could see clients seated at tables with white linen tablecloths and attending waiters dressed in threadbare dinner jackets. Faded elegance of times past.

Pitar Moș, was an easy street to find. A brass plate announcing 'Facultatea de Limbi și Literaturi Străine', Faculty of Foreign Languages and Literatures. The door was locked so no visit was possible but the thought of teaching there excited me.

At a pedestrian crossing I crossed the boulevard and began retracing my steps home. It was lunch time and I found a restaurant on a corner. On entering, I could see that the dining room had been elegant. Despite a dearth of diners, I had difficulty attracting the waiter's attention. I ordered "moussaka" even though I'd never ever heard the word before. After a wait, he put a plate down in front of me. It smelled delicious and I ate it with relish without the slightest idea what it was. I'd never before eaten aubergine. After that experience, I enjoyed solitary lunches there occasionally. Over time, I learned the entire menu, savouring everything.

One of the Romanian guards barred my way by raising his rifle across his chest when I presented myself at the British Embassy the following day. As I was about to explain who I was in Romanian words studied earlier that morning for this very purpose, a self-assured, well-built man in his early 50s wearing British clothes approached the gate from inside the embassy compound and spoke to the guard in Romanian. The guard lowered his rifle. Immediately, I recognised this man's function. He had the same confident bearing, stature and the all-seeing, faintly amused eyes of the Black Watch sergeant who had trained us in weaponry with the 3rd Battalion Gordon Highlanders in Aberdeen.

"Welcome to Bucharest, Mr. Mackay! I'm Bob. The Cultural Attaché is expecting you." Bob was the take-charge kind of man I immediately felt comfortable with, the kind of competent individual I'd met in the Warrant Officers mess in many a military camp.

The Cultural Attaché, Tony Mann, made me welcome. For an hour, he explained what I needed to know. He was to be my personal contact in the Embassy. He was an officer of the British Council but the British Council was not permitted to operate in Romania and so he bore the title of Cultural Attaché. It would be better if I did not mention the British Council during my stay. He summarised the troubled history behind his warning.

In 1949, the Romanian Communist Party expelled British diplomats. The Embassy's Romanian employees, all those who had worked for the British Council and for the Information Office, were accused of espionage and treason and condemned to 25 years hard labour.

In 1961 the Foreign Office became impatient with the British Council's determination to have its former staff released from prison. Britain and Romania signed a cultural agreement in 1962. A handful of prisoners were released but the British Council still considered the matter of its former employees unresolved.

Tony explained that although his work was encouraged by the Foreign Office given its hunger for reciprocal trade relations with Romania, his work was hampered by the Romanian Government. He would appreciate any help I could give him by dealing myself directly with Romanian officials and so reducing his need to battle with a devious Romanian bureaucracy.

I appreciated learning about the intrigues between the British Embassy and the Romanian Communist Party. I especially appreciated the free hand I was being given. He was also warning me that I was an employee of Bucharest University, *not* of the Embassy. Without diplomatic status I was unprotected and vulnerable. His next comment made my precarious situation clear.

"Please leave your apartment tomorrow at nine in the morning and don't return until eleven. Clear?" I nodded. "Our specialists will sweep your apartment and phone for bugs."

"They'll deactivate them?"

"No! We cannot do that. But we like to know what kind of technology is being used."

In London, the Foreign Office had forewarned that I would be under surveillance 24 hours a day. I was now behind the Iron Curtain where everybody, Romanians and Westerners alike, were under suspicion and constant scrutiny.

Tony ran through a series of warnings and offers of cooperation. The list was long. Anything, *anything at all*, I could take to him and he would help find a resolution. However, as a non-diplomat working at the University, it would serve me well not to be identified with the Embassy. The Communist Party created a law that forbade Romanians from associating with Westerners. The Romanian Secret Police, the dreaded *Securitate*, ran a vast network of agents and informers. Anybody who demonstrated a willingness to associate with me was likely working for the Secret Police. If I initiated an association with a Romanian, I could endanger them. I should assume that I was constantly being watched.

Tony had arranged to take me home for tea to meet his delightful wife Sheila. He had also invited an English engineering exchange student, Peter, who had only one week left of his six-month stay. When Peter heard I was a walker, he offered to take me to the mountains above the Prahova Valley the following Saturday.

Bucharest's elegant University Square. It was a pleasure to walk by here every morning.

3
HOW ALL THIS BEGAN

The Post of Visiting Professor

Only a few months earlier, I had been an MA student at Kings College, Aberdeen University. Before sitting my finals, I had seen a job advertisement placed by the British Council in a national British newspaper: "Visiting British Professor of Phonetics at the School of Foreign Languages and Literatures of Bucharest University, Romania". As a twenty-four-year-old who had worked on leaving school in Dundee, Scotland at eighteen, in the banana plantations on the island of Tenerife and travelled extensively in several European countries as well as Morocco, I decided to offer myself as a candidate for this romantic post.

A selection committee made up of members of the British Council, the Foreign Office and various British security services had interviewed me and offered me the position. The chair of the committee had made it clear that I was not the best qualified candidate, academically. Others had postgraduate degrees in phonetics, the Scottish MA was an undergraduate degree.

The chairman had been blunt. "This Bucharest post is a challenging one. Ceaușescu's Romania is oppressive. The Communist Party forbids

Romanians to socialise with Westerners. You must be highly self-sufficient to survive this lonely post. Your past work and independent travel experiences suggest that you have what it takes." Committee members had nodded soberly. At my security briefing at the Foreign Office, the officer had been even more frank.

"The Romanian Secret Police will assume that you are a British agent and treat you accordingly. Your apartment will be bugged, your telephone monitored. Their agents will have you under permanent surveillance. To coerce you into working for them, they may try to compromise you. They can accommodate any sexual preference. If you succumb, they may threaten to divulge your misdeeds unless you cooperate. The British Government forbids you from publishing anything about your stay in Romania until five years have passed."

Discovering the Romanian Language

None of my professors in Aberdeen could help me with the Romanian Language. "Your best bet is Edinburgh University," one told me.

I wrote to Edinburgh University. When no reply came, I hitch-hiked the 150 miles and presented myself at the desk of the receptionist to the Department of Phonetics. On learning that I'd be spending a year behind the Iron Curtain, she offered me some motherly advice.

"Och, son, dinna trust onybody ahent the Iron Curtain. And mind they communists! Given half a chance, they'll lock ye up an' throw away the key!"

Having duly thanked her for her help and her graphic warning, I spent the morning listening and re-listening to samples of Romanian speech in the phonetics listening laboratory. Most useful was the rendering into clearly enunciated Romanian of the well-known standard phonetic passage, the 'Story of the North Wind and the Sun'.

The North Wind and the Sun were disputing which was the stronger, when a traveller came along wrapped in a warm cloak. They agreed that the one who first succeeded in making the traveller take his cloak off should be considered stronger than the other. Then the North Wind blew as hard as he could, but the more he blew the more closely did the traveller fold his cloak around him; and at last the North Wind gave up the attempt. Then the Sun shined out warmly, and immediately the traveller took off his cloak. And so, the North Wind was obliged to confess that the Sun was the stronger of the two.

Within hours, I could recite from memory, the entire passage in Romanian and produce even those peculiar speech sounds that were not used in English.

Exhausted, I took a short break. It wasn't exactly a lunchbreak since I had little or no money. On my way back, I nodded respectfully to a craggy, unsmiling man.

"Stop! Do I know you?"

"No, Sir. My name is Ronald Mackay. I've just completed my MA at Aberdeen. I'm here to learn the sounds of spoken Romanian. I'm going to teach phonetics at Bucharest University."

"Bucharest? Hungary?"

"Bucharest is the capital of Romania, Sir."

He dismissed the rectification. "You? Teach phonetics!" He was horrified. "Nobody in Aberdeen University knows anything about phonetics."

I wondered at this Sassenach's arrogance.

"I am Professor Iles!" He loomed over me. "Head of the Department of Phonetics here in Edinburgh."

Before I could compliment him on how useful his Romanian tape was to me, he wheeled me round 180 degrees. "Out! I cannot have just *anybody* wandering in at will to use resources reserved for *my* graduate students."

Too diminished by his discourtesy and too crestfallen to let him know I'd made a formal request to use the facility, I left without a backward look. So much for my introduction to the Romanian language!

Are you Fit to Serve behind the Iron Curtain?

The terms of my appointment demanded I undergo a thorough medical examination conducted by a Government-appointed doctor in Aberdeen. The venerable doctor ordered me to strip to the waist, then probed with cold fingers and listened using an icy stethoscope. He looked at my tongue, hammered my knees then pointed to his scales.

"Ten!" He scratched his number into the form he'd been provided with by the British Foreign Office before running a wooden level to the top of my head, "Seven!" More scratching.

"Seven?" I queried. I knew that his "Ten" was stones – 140 pounds.

"Five foot seven."

"I'm five-seven-and-a-half," I ventured. That half-inch made all the difference to me.

"Are ye tellin' me my measure's no accurate enough for ye?" He was annoyed.

I decided that the half-inch was irrelevant. He scratched 5'7" on the form.

Finally, he wrapped a heavy cuff around my upper arm, pumped air into it and just a moment before my blood circulation was cut off, he released the pressure. He shielded the form with his left hand. "This information is for Government eyes only! Confidential!"

I could see and read the final question: *"Does the candidate show the mental stability necessary to serve in a British Government hardship post overseas?"* To avoid running any risk of his scratching in a *"No!"* I obediently looked away and did my best to appear *compos mentis*.

He asked about my parents, siblings, my experience at Aberdeen University and then scratched something about my mental state into the final box. Dutifully, I kept my eyes averted. Finally, he read the last line: "Country of Appointment: Romania".

For the first time, he looked at me with respect. "Romania?"

I nodded.

"Romania! That's a Communist country!" I nodded, expressionless.

"And what might ye be telling doin' behind the Iron Curtain?" The old doctor was awed, curious.

I met his eyes, paused and lowered my voice, "Government information only, Doctor. Confidential."

"Of coorse! I wasna thinkin'. I understand! Excuse the question, will ye, Sir? Confidential! Of coorse!" Flustered, he hurriedly scratched his signature, folded the form, and placed it in the stamped, self-addressed envelope disguising his embarrassment. "I'll post it this very night." He stood and offered me his hand, "Guid luck! Mind, noo, an' come back safe!"

In his eyes I was no longer a callow, 140-pound, 5'7" youth; I was key player in the intrigues of the Cold War.

Like the Scots, Romanians take pride in their wide array of national dress.

4

TRAVEL WARNING

The following day, I presented myself at the Rector's office, the administrative centre of Bucharest University. The unsmiling officer responsible for my stay led me to a wood-panelled room with a highly polished table surrounded by heavy ornate chairs. Previous gowned Rectors regarded me gravely from the walls. Seating herself on a carved wooden throne she gestured to one of the more uncomfortable chairs and regarded me sternly.

"I am appointed by the Rector to oversee all non-academic matters for the period of your appointment." Excellent English. She slid me a blue cardboard University ID. The card attested that I was the *Profesor Invitat la Universitatea din București*. It bore my passport number and date of birth, proof that I officially existed in Romania.

My apartment belonged to the university. She slid an inventory of the apartment's contents across the table. My teaching schedule would be assigned by the Head of the English Department, my academic superior, Madame Ana Cartianu. My salary would be paid monthly in Romanian Lei. I was forbidden to take Lei out of the country. I must collect my salary on the last day of every month at the Communist Party Base in the building where I taught on Pitar Moș.

She paused, then even more sternly: "You are free to travel within the city of Bucharest. If you wish to leave the city, you must ask permission from the Rector's office, in writing, two weeks in advance. You must state your proposed destination, purpose, route, and duration of your trip. If permission is granted," she stressed the conditional "it will be in the form of a permit for that specific trip and that trip only."

I didn't like what I was hearing.

"You must carry the permit with you for the entire period out of the city, show it to any official when asked and surrender it to me immediately upon your return. Under no circumstances may you visit the oilfields in the Plain of Wallachia." She showed me these on a classroom-style map.

The official welcome, if welcome it could be called, was over and I was out in a street bright with autumn sunshine.

"I've just been issued a verbal order that limits my freedom of movement!" I thought to myself. The planned trip to the Carpathians this weekend was out. Spontaneous trips were prohibited. Perhaps all trips might be prohibited if, according to Tony, written permission was likely to be intentionally delayed. "This amounts to permanent quarantine!" The injustice dismayed me! I sat in a café, thinking. Had I come all the way to Romania just to be stuck in Bucharest, beautiful though it might be, for an entire year? A solution dawned on me.

The unsmiling woman in the Rector's office was telling me that although I had the same professional obligations as my Romanian colleagues in the University, I was to enjoy fewer rights when it came to travelling because I was a foreigner from the West. She was demanding I restrict my own travel to save the State the trouble. If I complied, I would be doing the State's job for it! *"What if,"* I asked myself, *"what if I take the view that responsibility for enforcement falls on the Romanian State not on me?"* Perhaps the State didn't even possess the capability to monitor my every movement. There was one sure way to find out. There and then I decided that I would simply ignore her warning. "After all, what can they do to me?"

The truth was, I had no idea what the State could do to me!

And so as planned, that Saturday I met Peter at the Gara de Nord where I'd arrived only a few days earlier. He explained the simple process involved in buying tickets to the village of Bușteni in the Prahova Valley. We bought return tickets for the same day. His imminent return to the UK prevented him from spending the entire weekend in the mountains.

On the train, he asked me if the Rectoria had placed any restrictions on my travel.

"Yes," I told him, "but I've decided it's up to them, not me, to enforce their rules! This trip is a test of their capabilities. They either stop me or they don't!"

He laughed. "I made the same decision when I arrived! Any time I've been asked for ID outside Bucharest, I show my University card. I've never suffered any dire consequences!"

We were two of a kind. I was sorry that he was about to return home to the UK.

Casa Scinteii, the state publishing house of Romania and emblem of the communist regime.

5

THE PRAHOVA VALLEY

First Visit

Along with a cluster of Romanian hikers, men and women ranging in age from their early 20s to their 60s, Peter and I descended from the train to the bracing mountain air of the Carpathians. The Prahova, a fast-flowing mountain stream, separated the Buçegi Mountains that rose above the west bank from the Baiu range on the east. Scattered about were wooden houses with white picket fences and larger and beautiful villas built from natural wood. Peter told me that these villas had been requisitioned by the State in 1949 when private property was abolished and now served Romanian tourists from Bucharest year-round, hikers in summer, skiers in winter. Some were reserved for members of the Communist Party, others popular with the proletariat and still others with the intelligentsia. The villagers wore brighter colours than the peasants in Bucharest, the backs of their waistcoats often beautifully embroidered. I felt that it was easier to breathe out here in the countryside.

Peter pointed out a prominent cross atop a peak well above us. "The Heroes' Cross, on the summit of Caraiman. Twenty-three hundred metres. Erected to honour those who died in the First World

War." That's where we were heading. From there we would be able to see the Cabana Caraiman and the Cabana Babele – two of the many hostels where, on future trips I would spend the night.

In single file, with small groups of hikers ahead and behind but none daring to speak to us, we set off up the steep path through the woods. After hours of slogging we emerged onto the high plateau close to the *Crucea Eroilor*, the Cross. We took some photographs. Without a word, the Romanian hikers headed on towards one or other of the two hostels.

"They dare not speak to Westerners!" Peter shrugged.

I decided I would learn Romanian, one step in bridging the gap.

We ate our sandwiches while enjoying the magnificent views of interlocking mountains and valleys, forests and rock faces that Caraiman offered us. It was evening by the time we descended again into the valley and boarded the train for Bucharest. Peter had been a good companion. He'd given me a first insight into train travel and hill-walking in Romania. I'd learned a great deal from him in that single day.

First Solo Trip to the Bucegi Plateau

During the following week I stayed in the city and explored alone. Madame Cartianu was supposed to schedule an orientation meeting and details of my teaching schedule. Classes were due to begin shortly. When still no summons had come by Friday, I decided to go off to the Carpathians again. I would take the train to Sinaia and follow a different, longer route into the mountains and stay two nights at either Babele or Caraiman.

In the Gara de Nord, I bought my tickets using pre-prepared Romanian words and boarded the train. I was one of the first and selected an empty carriage. New passengers would examine my empty carriage from the corridor, observe me in my British clothes and move on. The result was that I had the carriage to myself all the way to Sinaia.

When the train stopped in that picturesque resort village, all I had to do was follow others dressed as hikers, cross first the bridge over the river then the road that led north to Transylvania, enter the forest and begin the steep trudge upwards on the trail. A score of hikers in silent pairs or small groups, men and women ranging from their 20s to their 60s. All were dressed more appropriately for the mountains than I was with stout hiking boots, pant-legs stuffed into woollen socks. Some dressed in plus-fours of the kind worn by my grandfather. Their canvas jackets showed wear. Strapped to well-used, rucksacks were warm woollen sweaters. The older walkers carried walking stout poles, some with carved handgrips.

With the briefest of greetings to one other and to me, each group established its own pace. They had no intention of forming broader alliances. That suited me since as I understood little Romanian other than what I could guess at from my grasp of Latin and could speak less. I conserved my energy for the steep trail through the magnificent beech and chestnut trees and the sweet-smelling pines. It was not unlike climbing the path to the summit of Ben Nevis but with the added beauty of trees that thinned out and eventually disappeared as we ascended.

Soon, I was outstripped, on purpose, it seemed. Understanding and so unoffended, I settled in some 500 yards behind the others. I was young, fit from manual labour and military service, and used to hiking in Scotland, but had no proper equipment. I wore no boots, having been able to afford, before I left, only a pair of used but stout shoes from an ex-army store in Aberdeen. Most of my clothing during my student years had come from second-hand from the same store.

As we left the woods and approached the plateau the path steepened. Tumbling streams created steep valleys and the path twisted to avoid banks of scree too dangerous to walk on. The views of the valley and the mountains to the east were glorious. I climbed easily.

Once on the plateau, the various groups stood together, drinking from water flasks and enjoying the panorama. They were more relaxed up there. Without asking me any personal questions, they proudly identified distant peaks to the east and the west as well as the neat villages nestling in the valley far below.

Closer but still separately, we tramped across the plateau past Omul, at 2,500 metres the highest mountain in the range. The hikers appeared to be experienced and intimately familiar with this walk. They talked quietly within their own groups.

Soon, separated by several kilometres, both hostels, Babele and Caraiman, could be seen. The band split up, each group heading towards its preferred cabin. I paused to give them a good start and then followed the smaller band, heading for Caraiman. I reckoned I stood more chance of finding a bed in the hostel with fewer people.

The hostel was a fine old cabin of weathered wood, a far cry from the stone cottages and converted outbuildings that I knew as hostels in Scotland. Those who'd preceded me had informed the warden of my arrival. He was young and had the air of a country doctor on holiday. Discreetly barring my way, he asked me who I was, where I came from and what I was doing up here in the Carpathians. I explained as best I could, drawing on both my Spanish and – with greater effect -- on my Latin.

Patiently, he explained that besides being unable to admit me without a reservation, hostels were for the exclusive use of Romanians. Other hikers pretended not to eavesdrop.

Determined to spend the night, I drew on all my linguistic and some more creative resources. I fibbed that Caraiman was the specific hostel where the Rector's office had suggested I spend the night. I showed him my impressive university ID card and pointed to the official signature of the Rector himself.

Now the warden was in a quandary. As if to share his burden, he invited others to inspect my ID. Curious, they examined the blue card saying nothing. They were clearly uncomfortable yet impressed by a Westerner whose identification proved he was a visiting university professor. To the warden's relief, an older gentleman accompanied by his wife, supported my case.

"He is a professor at Bucharest University. He is entitled to stay."

That resolved the impasse. I'd won my first battle or rather a kind Romanian gentleman had won it for me. There would be many, many more battles to fight.

Relieved that someone else had made the decision, the warden had me sign the register and pointed to a dormitory. I found my bed number, lay down, exhausted, and closed my eyes. When I opened them, I saw one of the women hikers strip off and make for the showers. Alarmed that I'd entered the women's dormitory by mistake, I rushed to the warden to correct my error. He looked at me amused. *Didn't I know that hostel dormitories were mixed? Did I have a problem with that?* I was taken aback. Hostel dormitories in Scotland were segregated. Embarrassed, I assured him I had nothing against sharing quarters with members of the opposite sex. And from that day on, I enjoyed mixed dormitories and the simple delights they afforded.

The hostel fee included a satisfying dinner made by the warden himself. Most hikers avoided me but, dinner over, the older gentleman and his wife joined me. Both spoke English. In a matter-of-fact way, he summarised my situation as a visitor from the West.

"All Romanian adults are assigned a *'base'* by the Communist Party. Their personal files are maintained in that *'base'*. The Communist Party forbids association with Westerners. If for whatever reason a Romanian citizen has contact with a Westerner, he or she must report the conversation in detail to the Party official at the *'base'* where the incident will be recorded. Every Romanian strives to keep their file to a minimum. The less contact we have with foreigners the simpler our lives. Your university colleagues, your neighbours, anybody with whom you might have chance encounters, will strive to avoid you. We Romanians do our best to avoid a visit to the Party 'base' because an incriminating record can be used to deny a future post or trigger a security check."

I was taken aback.

"Secret Police informants are everywhere. Traps are set. If you plan to continue your walking excursions, it will serve you well to dress in Romanian clothes and use a Romanian backpack. Until you master the language, speak little or not at all. If you have studied Latin, you may learn the language quickly. Once you do, you may be mistaken for one of Romania's minorities – a Hungarian, a Saxon, a Schwaab, a Bulgar, a Tartar or a Moldovan."

When he had finished, he politely informed me that he and his wife wished me well. Curious others had been watching. Raising his voice so that others could hear, he said:

"On Monday, I will report our conversation to my Communist Party 'base' in Bucharest, with the accuracy a mind as old as mine is capable of." They rose with dignity and shook my hand. The curious watchers dispersed. I received only smiles and nods from then on.

However, following his advice, I decided to buy Romanian clothes and walking equipment, to learn Romanian, and to continue my excursions into the Carpathians and, if I could, explore the rest of this fascinating, beautiful yet mysterious country.

Hiking in the Carpathian Mountains was a great pleasure. My work schedule gave me lots of opportunity.

6
GETTING DOWN TO WORK

Eleven Red Roses

Only days before the teaching semester was due to begin, the Head of Department, Madame Cartianu had still not contacted me. I'd been advised to arrive three weeks early to give her time to assign my workload and for me to prepare.

"Visit her in her home," Tony Mann suggested. "It's not strictly correct, but you don't seem to be one to stand on ceremony." I had told him that I planned to leave the city whenever I chose to and leave the rest up to the authorities. "I didn't hear that," had been his reply.

Tony gave me Madame Cartianu's home address and so the following afternoon I pressed and donned my only suit, took a trolleybus to the Piața Romana, entered a florist's shop and used the phrase I'd researched to ask for a dozen red roses.

"Un cadou pentru o domnişoara?" Inquired the middle-aged saleswoman with a flirtatious smile. "A present for a lady?"

"Da." "Yes."

"In Romania, a gentleman presents eleven roses to a lady. By doing so, he signifies that she, the most beautiful bloom of all, completes the dozen."

Her look suggested I could be in for quite the surprise were I to present *her* with eleven roses. I thanked her. She carefully wrapped them, smiling warmly.

"Before presenting them you must tear the wrapping paper thus!" She showed me where and how carefully to tear. "Roses presented to a lady must be exposed!" While exaggerating the word "exposed", she mimicked the act of disrobing.

Her worldly charm delighted me. Since that day, I have, with success, followed her advice.

I found first the street and then the elegant-but-fading 1920s apartment building with an ornate wrought-iron entrance. Madame Cartianu's apartment was on the first floor. I mounted a wide, ill-lit, marble staircase and once on the landing, paused to straighten my tie, I knocked, bruising my knuckles on the solid portal. No response! As my eyes adjusted to the gloom, I noticed a huge iron knocker and used it to thunderous effect. Then the door was unlocked and cracked open an inch to show the white of a woman's eye and part of her cheek.

"Ce vrei?" "What do you want?" The suspicious Eye appraises me.

"Numele meu este Ronald Mackay," With formality, I announce myself. In prepared Romanian I tell the Eye that I am the British Exchange Professor come to pay my respects to Madame Ana Cartianu. The Eye is silent. The door begins to close.

"With roses!" I rip the wrapping paper and extend the blooms. The door opens just wide enough for the roses to be taken from my hand.

"Please wait, Sir." The door closes. I stand in the gloom. The door reopens, wider this time, to show a woman who looks so old, frail, worn, and utterly drained that I'm taken aback. Is she in mourning? Recuperating from an illness? I feel like an inopportune intruder.

"I am Madame Cartianu." Perfect English. An unsuccessful attempt at a smile. She gestures me inside. The apartment is grand but closed doors and heavy curtains make it gloomy. She leads me into a dark sitting room lined with bookshelves.

"I am unable to spend time with you." No apology. No invitation to sit.

I cover my bewilderment by begging her pardon for arriving unannounced. "I want only to pay my respects and to learn about my teaching duties so that I can be adequately prepared." Wordlessly, she inspects me. What she sees brings no light to her day. I'm conscious that my suit is second-hand, and my shoes worn. However, it consoled me that I'd pressed the suit and polished the shoes out of respect for her position.

Thankless, she holds the eleven roses in an ungrateful hand. I wish I'd brought the full dozen.

"Where are you from?" Her question puzzles me.

"The UK"

"The UK?"

"Coupar Angus, a village in Scotland. School in Dundee. MA from Aberdeen University."

"You are English?"

What kind of an Englishman comes from Scotland? I think to myself. "I'm Scottish." Madame Cartianu is not impressed.

"I will arrange for you to be briefed by the Dean. You may leave now."

I will the eleven roses to wither in her hand. She leads me back to the front door, opens it and gestures me through. I hear the key turn firmly in the lock.

Should I, perhaps, return to the charming florist and ask her where I have gone wrong? Instead, I walk home feeling dispirited for the first time since arriving in Romania.

Dean Ion Preda

The following morning the telephone rang in my apartment.

"Mr. Mackay, my name is Ion Preda. I am Dean of the Faculty of Foreign Languages. Will you do me the honour of having lunch with me today?" Immediately, I warmed to the friendly timbre of his voice and his impeccable English accent.

"I will be delighted to have lunch with you, Professor Preda." After three weeks of alsmot total isolation this would be my first official contact with the university.

As agreed, I turned up at the dining room of the Lido Hotel. I already knew the hotel from my walking excursions throughout the city. Inconspicuous on the main Magheru Boulevard, the Lido possessed a faded elegance.

Ion Preda, in good quality if well-worn flannels, sports-jacket and tie, colours matching, turned out to be a delightful host and conversationalist. In his early 50's, he radiated education and breeding and possessed the same courteous air as the gentleman who had intervened on my behalf at Caraiman. Smiling a welcome, he led me to an elegant dining room. The formally dressed *maitre d'hôtel* showed us to a table set with linen, silverware, and sparkling glasses. Gone was the disappointment of the previous day.

Ion invited me to call him by his Christian name. He conveyed Madame Cartianu's regrets for not briefing me herself. I was happy.

The menus were in Romanian of course and so I told Ion I would have exactly what he was ordering.

We started with a steaming *ciorbă*, a delicious Romanian soup. On the expanse of white tablecloth between us, the waiter had placed a plate containing five or six thin green pods. We began our meal. Ion picked up one of the pods, bit and chewed. I confidently did the same but unwittingly took a bigger bite. For a moment I could taste nothing at all and then my mouth and nasal passages came under attack. My eyes teared, my brow poured sweat. I lost track of what Ion was saying, able only to focus on drawing brief, shallow breaths. Ion looked at me with concern.

"My dear man! I should have warned you! You have probably never before eaten *ardei iute* – hot chili peppers!"

I had never even *heard* of hot chili peppers. The waiter escort me to the men's cloakroom where I rinsed my mouth and face in cold water, fractionally reducing my suffering.

Back at table, I assured Ion that I was fine, but I let him talk for the rest of the meal.

"How would you feel about teaching semantics?"

Semantics? I thought, the study of how words and phrases carry intended meaning and so allows speakers and listeners to understand one another. I was puzzled. I reminded him that I held the post of Exchange Professor in Phonetics.

"Of course, of course," he agreed. "Nevertheless, Madame Cartianu has decided that before she retires, she would like to teach the phonetics course herself. She believes you will have no objection to her pulling rank on you?" Still semi-incapacitated, I confirmed I had no objections. And so, my teaching duties were to be in the field of semantics, not phonetics.

Only later did I discover that Madame Cartianu had been horrified when she had heard my Scottish accent. At once, she had decided that she would not compromise the standards of her department by exposing students to my spoken English. Her reaction may seem odd to native speakers of English, but it was a decision based on well-founded historical principles.

Any British man or woman can identify what region of the British Isles a speaker comes from by their accent. The English, Scots, Irish and the Welsh all sound different. Even within each of these regions, accents vary.

When the phonetician Daniel Jones decided to describe in detail how English is pronounced, he first had to decide what *accent* he was going to select for his study. The obvious choice would be the '*best*' accent. In most countries the most prestigious accent is that spoken in the capital city, for example the French spoken in Paris not in Marseilles, the Spanish spoken in Madrid not in Seville. However, the London accent does not carry the greatest prestige in the UK. That honour is reserved for the one and only accent that does *not* signal the speaker's regional attachment but how they were schooled.

Those who attend private schools, confusingly called *public schools* in Britain, learn to speak with a non-regional accent that Daniel Jones called Received Pronunciation. 'BBC English', the 'Queen's English' and 'Oxford English' are less technical terms. Accordingly, Daniel Jones devoted his attention to Received

Pronunciation, the prestige, non-regional, British English accent and as a result it had become the standard most Europeans aimed at.

I began to plan quite a different course for the students I hadn't yet met.

Colleagues?

Ion Preda invited me to the School of Foreign Languages to meet my new colleagues and learn the layout of the building.

He greeted me at the front door. At the end of a dark marble hall, he ushered me into a high-ceilinged room well-lit by casement windows and lined with glass bookcases, all entirely empty. A dozen men and women, mostly standing, talked in pairs. As Ion and I entered, conversation ceased, all rose, and, one by one, Ion introduced me to my colleagues.

The first was a dark-haired man in his early 40s with cautious eyes, "Chițoran!" He gave me his surname, his hand, and a penetrating look. Later, I learned that in addition to his being a teacher and senior administrator, he managed the Communist Party 'base' that served the School of Foreign Languages.

Next came two kindly, scholarly-looking gentlemen, Professor Levițki and Professor Duțescu, both Shakespeare scholars. I would rarely see one except in the company of the other. The next, in his early 60s, widened his fixed smile and bobbed his head, "Professor Ştefanescu Draganeşti."

'What a beautiful name,' I thought, *Ştefanescu Draganeşti!*'

'We are neighbours!" He offered to introduce me to travel on the public transport system.

"In the three weeks since I arrived," I smiled to all, "I've used your trolleybuses and trams as well as *shanks-pony* to explore much of your beautiful city."

The idiom drew laughter from the group and broke the formality. I would find that there was not a single English idiom that my colleagues were not familiar with. The fact that the implication in my phrase, *"In the three weeks since I arrived,"* registered not at all with any of them, reminded me that initiating contact was beyond the privileges their ever-watching State permitted. Immediately, I felt ashamed I'd insinuated that I had felt neglected.

A younger, slighter man shook my hand and greeted me like a long-lost friend, "I am so awfully pleased to make your acquaintance, old boy!" A perfect imitation of Bertie Wooster, P.G. Wodehouse's archetypical English gentleman. This was Adrian Nicolescu's normal accent and natural upper-class manner. Whenever we met in future, he would greet me with a smile, an extended hand and, "So pleased to see you again, dear boy. It's been absolutely ages!" All he lacked was a monocle! Later, I learned that he came from a once well-to-do family and, as a young man, had been educated by an governess, herself a minor member of the British aristocracy.

Next to introduce himself was: "Andrei Bantaş, lexicographer! I have compiled and published a Romanian-English and an English-Romanian dictionary, Professor Mackay. I will see that you receive copies." I thanked him.

Finally, several timid, women professors, all a little older than me. I saw few of them again perhaps because of our different teaching schedules.

My colleagues appeared alert, intelligent and to project an air of heightened vigilance. Their English was perfect; indeed, so perfect that I wondered if English was not their mother tongue. I discovered that only a few of the older ones had ever been outside Romania. They showed great respect, even warmth, for Professor Preda while with Professor Chiţoran they appeared deferential and uneasy.

None tried to engage me in small talk. None asked me any questions. None asked me when I'd arrived or what I thought of Bucharest. None, other than Professor Ştefanescu Draganeşti offered me any personal assistance to settle in. I realised that I was witnessing the reality behind the warnings given to me by the Foreign Office and later by the Cultural Attaché. This was indeed, a police state.

In the days, months, and years to come, I was to find that whenever I entered the common room, all conversation stopped. Those present would greet me but further conversation was not encouraged and, as if to forestall even that possibility, my colleagues would quietly slip away. Within a few weeks I would come to appreciate that my presence so inhibited my colleagues that it was best I use the common room as infrequently as possible.

In passing, on our way to the teaching classrooms on the floors above, Ion pointed out the Communist Party 'base'. I would have no reason to visit the '*base*', he told me. When I mentioned that the Rector's office had told me I would collect my salary there, he said he would arrange for me to collect my salary in the faculty common room with nobody else present.

The seminar room I would teach in was on the first floor and almost identical to those in Aberdeen University, furnished with a long table flanked by a score of wooden chairs, the professor's chair at the head. Tall bright windows looked onto the street below, Pitar Moş. A chalk blackboard extended the length of one wall.

"Classes will begin on Tuesday morning at eight." Ion shook my hand. I was free to go.

My duties were light, only four sessions each lasting two hours, Tuesday to Thursday. That gave me a very long weekend. Ideal! Then again, I reflected, the British Council could hardly demand more for the meagre £30 they were depositing each month into my British bank account. I felt greater appreciation to the Romanian authorities who provided my rent-free apartment and a monthly salary in Romanian Lei. Because Lei could be neither taken out of the country nor converted into hard currency, my entire salary would return to the Romanian Government.

My Students

My students, all in their first year of studies for a degree in English language and literature, were an absolute delight. They were intelligent, enthusiastic, highly motivated, courteous, hard-working, and appreciative of my modest repertoire of skills and knowledge. Their curiosity and conscientiousness matched or exceeded that of the British students I'd known. They reciprocated my esteem principally, perhaps, for my novelty value as a native English speaker from the glamorous West, the only one they were likely to meet.

A bell announced the fifteen-minute break within each two-hour session for all classes, so the corridors and staircases were crowded. Students other than my own would look at me with open curiosity but none, not even those in my class, would approach me. If I sought refuge in the common room my presence embarrassed my colleagues. The solution was to remain in my classroom where my students had fewer inhibitions about talking to me.

Students in my classes were mainly young women. During the break, they shed their competitiveness and showed curiosity about life in the UK and to my initial embarrassment, delighted in telling indelicate jokes. The more explicit, the more they laughed. They appeared more worldly than the average young person I was familiar with in Scotland. Nothing embarrassed them; on the contrary they took delight if they succeeded in causing me embarrassment. They had little difficulty in mortifying me given the innocent social customs of the youth I had experienced in Scotland in the '40s, '50s. As a student I had immersed myself in studies during term-time and physical work during the vacations to pay my expenses, I'd had neither time nor money for a social life and none of either for frivolity. My Romanian students' joyous, titillating conversations and explicit jokes made me appreciate that it was I, not them, who had lived behind a protective curtain, at least as far as sex was concerned.

"Why did Aristotle Onassis hold up his hands like this to Jackie Kennedy on their wedding night?" The girls held their hands up palms to me, fingers wiggling.

"I've no idea," was my innocent reply.

"So that she herself could choose the finger."

This prudish Scot looked at them, face blank. Then the penny dropped. I flushed a deep red. Their joke had caressed not one, but two sensitive points and they hooted in double triumph. For the remaining hour of class I avoided eye contact with them.

They avoided politics, East or West, as did I.

My two-hour classes with such stimulating, entertaining and

thoroughly admirable students were the only regular contact I had with Romanians during my early months. Classes were the highlight of my week despite the occasional embarrassment.

I say *almost* the only contact because there was Karen. Karen was not a student. Karen had sought me out in the busy corridor during the first week of teaching before I decided it was better to spend the break in the company of my own students rather than in enforced segregation.

A proud Romanian in national dress.

7

WHO'S WHO?

Karen

"Ronald George Mackay?" The confident grey eyes looked directly into mine. They belonged to a most attractive, business-like young woman in her mid-twenties. Her dress and manner suggested she was neither professor nor student. Her natural charms were enhanced by a well-fitting tailored skirt and jacket.

Despite a bare month in the country, I'd become used to the feeling that in the initial microsecond of a first meeting, the Romanian had somehow taken charge and was capable of predicting my every reaction. This young woman, bold yet inviting, projected self-assurance.

"Professor Mackay, my name is Karen." She paused confident that I liked what I saw. "I am a graduate of Bucharest University. I have a dictionary that belongs to Professor Martin Murrell. Would you be so kind as to return it to him when you return to England?" It was less a request than the certainty of my full cooperation.

My watching students were unable to disguise their curiosity. Who was this woman bold enough to accost the British professor during the break? The bell rang and students swiftly returned to their classrooms.

Mine claimed the right to be unashamedly curious and watched Karen and me from the doorway.

"Perhaps," I thought, "this is a perfect opportunity for me to wrench control of the situation from Karen's hands, and see what might happen."

"Right now, Miss, I'm teaching a class. Please meet me at the gate of the *Casa Universitarilor,* the faculty dining room in an hour's time."

A fractional narrowing of her eyes indicated that I had surprised her. Karen had expected me to devote my immediate attention to her alone. I returned to my seminar room rather pleased. My students couldn't contain themselves. "Is she your girl-friend?" "Does she love you?" "Did you know her before you came to Bucharest?" "Are you lovers?"

That my students thought I was capable of attracting such a woman flattered me. I'd never figured particularly among those upon whom Cupid smiled. As a student in Aberdeen, I'd had neither the time nor the money to socialize and women never made the first move. On the rare occasions I summoned up the courage, I was rebuffed, gently or otherwise. It didn't bother me. The Scotland I grew up in was strait-laced and strict. Relationships were comparatively chaste at least until a formal engagement was contracted. *Lovers* dwelt in Hollywood.

I couldn't get through the second half of my class fast enough. It wasn't just that Karen was the first woman to approach me directly. I had two other reasons. I couldn't imagine two people less likely to have shared a common interest in English dictionaries than my predecessor, Martin Murrell and Karen. I had met him briefly in London. He struck me as a man who seldom left his desk, although he must have since he had a wife and two children, but he was certainly not a man to risk losing a coveted dictionary. The more important reason was that Karen had addressed me by a name that was not mine. Although born 'Ronald George', I'd abandoned the 'George' years before in favour of 'Mackay'. At 18, I'd made the change legally. At no time since then had I used 'Ronald George Mackay'. My passport and British Council records were in the name of Ronald Mackay. It took painstaking research and then a blunder to address me as "Ronald George Mackay".

I'm not a fast thinker. Hours after a conversation, it may strike me what I *might* or *should* have said. Unaccountably, however, faced with Karen, my thinking processes had gone into over-drive. I'd shown no surprise at her error and I'd made sure that we would meet again. Romania was helping me grow.

Karen was waiting at the wrought iron gate to the Faculty Dining Room only slightly irked. She would not enter the Casa Universitarilor but suggested a nearby park. I saw that she carried nothing but a purse. Once we reached the park, she gave me its history and the historical event it commemorated. She spoke English fluently. She told me she'd graduated two years earlier and spoke German, Russian and English as well as Romanian. She loved Russian history. She knew Bucharest intimately.

"If you like, Professor Mackay, I can help you become familiar with our city." My heart leapt. No Romanian, to date, had been so forthcoming.

I reminded her that she'd mentioned a dictionary. "Unfortunately, I forgot to bring it, but if you care to meet me next week, I will give it to you." She smiled into my eyes.

For reasons which were not entirely clear to me, I agreed wholeheartedly. Satisfied, Karen walked off with the grace of a model, conscious that I was watching.

The following week we met in Cişmigiu Gardens. My original assessment of Karen's charms was confirmed. She was fractionally shorter than my 5'7", immaculately groomed, with a fine figure, an intelligent face, and a pleasant voice. She maintained her business-like manner.

Once more, she apologised for forgetting to bring the dictionary.

Karen and I began to meet every week. The matter of the dictionary never arose again; nor did I challenge her about the error with my name. We were both enjoying our excursions. It seemed wise not to provoke.

Karen was a delightful companion. She had graduated top of her class and now held a government post in *cultural affairs*. No elaboration. An urban person, she loved the city's historical monuments, art gallery and its beautifully maintained parks. I loved, but she scorned, the Museul Satului, a recreated village museum in a park made up of period buildings and artefacts from all regions of Romania. It brought to mind the way of life, the skills and trades of the countryside of my own early years. Karen explained that she was a member of the *intelligentsia* and above folksy museums of interest only to the *proletariat*. I was learning that there were hierarchical differences in Communist Romania.

My arrangement with Karen was almost ideal. She had confidently approached me and was comfortable with the arrangement. I had a lovely and intelligent companion. Our meetings were an oasis in my otherwise solitary life. Why should I care if she reported to the Communist Party 'base' in her ministry? Our conversations were innocuous. I had nothing to hide.

We continued to meet and spend Wednesday afternoons exploring Bucharest. Although I told her about my hikes into the Carpathians, I was relieved she never suggested accompanying me. Many walkers however seemed to travel as couples and shared beds in the hostels. Karen was an urban woman and did not strike me as one who enjoyed roughing it. I'd become used to planning my own itineraries and dividing my life both in and beyond Bucharest into sealed compartments. Becoming further involved with Karen would only complicate a life that I was managing well-enough to keep my head above water.

The hardwoods in Bucharest's boulevards and parks were now shedding their leaves and the sun was paling. Karen never mentioned her personal situation nor introduced me to a single friend. Occasionally she would pass a known face on the street. They would exchange "Good day!" but neither would stop. We were alone on our sightseeing excursions. She talked about history, architecture, and the paintings in the galleries, never about herself.

We would have dinner together in a place of her choosing. I found no pleasure eating alone in restaurants and so it was a pleasure to dine with an attractive, enigmatic woman in pleasant surroundings in a secretive and undeniably mysterious city, tucked two days train journey behind the Iron Curtain.

Her preferred restaurant was Capşa's. It was more elegant and more beautiful than any restaurant I had ever visited. To me it represented the peak of European sophistication. From the moment I first entered Capşa's, I was transported back to the dazzling days of the Austro-Hungarian Empire. A liveried footman met guests at the entrance and led them to the door of the dining room where the formally dressed *maître d'hôtel,* leather-bound menus in hand, inclined a solicitous welcome. The dining room was grand, the ceilings decorated, the tall windows curtained in red velvet and the lights created intimate spaces. The silverware was set on crisp white tablecloths. That everything was threadbare and slightly fatigued, only added to my thrill.

Capşa's was invariably busy. Most diners were much older than us, at least the men were. They would stare shamelessly at Karen and then somewhat contemptuously at me. Karen's perfect grooming matched the elegant fashion of the few women present. My flannels and jacket, or on occasion my suit, failed to meet the standards of the men. I was unconcerned.

"Capşa's," Karen told me, "is the restaurant of choice of senior functionaries in the Romanian Communist Party, visiting delegations from Warsaw Pact countries and East German spies." Romanians liked to believe that their country was the target of East German intelligence. For all I knew, it was.

Romanian women I found to be unlike any I had previously met.

They were naturally flirtatious. Students in other classes, even occasionally students in my own class would glance at me in the most provocative manner, look away and then glance back to see if I was impressed. Pleased and flattered, I never responded, partly out of inexperience, partly because I felt it inappropriate for a professor to flirt with students.

Initially, Karen was formal and business-like but then she began to flirt a little. She would take my hand in the park and raise her face for a simple kiss. Given the circumstances, I found this uncomplicated state of affairs satisfactory.

By mid-December 1967 I was feeling content with my life in Bucharest. I lived in a series of compartments that never overlapped. In the city from Tuesday to Thursday, I met my students. We enjoyed one other's company and furthered their already advanced knowledge of the English language. I walked to and from the university enjoying the early morning bustle. I shopped at the peasant market stalls in the Piața Unirii on my way home. Occasionally I would make side-trips into back streets and Orthodox churches, places Karen avoided.

I made and ate most of my meals in my apartment alone, enjoyed occasional dinner invitations from Embassy acquaintances and spent my Wednesday afternoons and evenings with Karen. Weekends, I took the train to the Prahova Valley and gloried in the forests and the mountains. I was satisfied, pleased with all aspects of my life and work in Romania.

Content I may have been, but Karen was becoming unsettled. Initially, she had shown no interest in my apartment, now she wanted to visit. I ignored her hints. One afternoon we were caught in a shower of rain. Both of us were soaked. She suggested we go to my place to dry off.

After we dried ourselves in separate rooms, Karen clearly intended to put the seal on our relationship. I hesitated, unwilling to complicate my life in ways my imagination could only exaggerate. For a second time, I made a split-second decision and forestalled Karen just before the point of no return.

"Karen?"

"Yes!"

"I think we should stop seeing each other."

She looked stunned, then burst into tears.

I was taken aback and felt slightly flattered. In previous years I may have shed a tear or two over a pretty girl but I'd never been the cause of any heartache. Now, judging by her tears, I was breaking the heart of the most beautiful woman I'd ever known. I tried consolation, my arms around her in as comradely a way as I could, then words calculated to comfort.

"Karen, you are a beautiful, well-educated woman. Lots of young men would give anything to have you as their girl-friend."

She pushed free from my embrace, stopped sobbing and looked at me with incredulity.

"You think it's you? It's not about you!"

She saw my puzzlement, drew a deep breath and explained to this dolt of a Scotsman, "I'm not crying about you!"

I had heard correctly but my mind was in low gear compared to hers.

"Listen! I'm member of the *intelligentsia*. I am ambitious. When I was sixteen, I was invited to become a member of the Communist Youth Party. That gave me benefits; holidays at nice summer and winter resorts, guaranteed admission to University. I enjoy living well."

I was taken aback by her vehement honesty. She continued.

"To do better, I must be invited to join the Communist Party. As a member I will be assigned good living accommodation, offered promotion, earn a better salary. Perhaps the opportunity to travel abroad."

Her candour astounded me. Then she dropped a bomb.

"I've reported on short term visitors for the Secret Police for several years. This year they entrusted me to report on you for the entire year. If you break off with me, I can't perform my assignment. Failure will count against me. It is not for *you*; it is for *me* that I cry."

"The dictionary was just an excuse?"

"It worked, didn't it?"

I nodded.

"Since you've been here for four months and have failed to find yourself a real girlfriend I felt I'd be doing you a favour."

My self-image crumbled. Karen saw me as a loser. She saw me as a Westerner in liberal Romania, surrounded by beautiful girls but due to some inexplicable deficiency I'd never managed to get close to one of them! Her sympathy for this flawed Scotsman was so great that she thought she'd do him the favour of becoming his lover, as well as his informer!

I was too astonished to laugh at the irony. A puritanical Scotsman afraid to embrace what Romania had to offer him and so causing offence to a beautiful secret agent by rejecting her gift and thereby imperilling her career plans!

Silently, I sat staring at the exasperated Karen. Expecting nothing from a coward like me, she recaptured the initiative, dried her eyes and returned to her business-like self.

"I have a proposal." No romance in her voice now. "A proposal that will suit us both."

"Proposal?" Like the smart Romanian she was, she was five steps ahead of me.

"If you stop seeing me, I will be disgraced. For you too, it will be no better. The Secret Police will assign another informant to report on you. You'll never discover who it is. But if you continue to see me until the end of your contract, I promise you two things."

I was still struggling to keep up with her.

"I promise never to report anything negative about you and when my assignment is over, I will never inconvenience you again. We both win!"

Her proposal made sense. I could have a lovely companion with whom to visit museums, dine with, and even appear in public with at concerts -- no strings attached. Karen would step closer to what she coveted, an invitation to join the Communist Party.

"I accept your proposal," I said. Karen smiled again.

I kept my side of the bargain and continued to meet Karen until my first-year contract expired. We enjoyed each other's company. When we went to the symphony together for the first time, we bumped into people either she or I knew, including a First Secretary at the British Embassy. The following day he issued me an urgent summons. Unsmiling, he led me to the Embassy *safe-room* swept clear of bugs.

"By appearing with a Romanian woman in public you are putting her in serious danger!" I listened to him scold me in polished public-school tones as if I were an errant schoolboy.

He finished his tirade. I thought for a moment and then told him the truth.

"Karen was the informant assigned to me by the Securitate. We have an understanding."

"How could you possibly know she's an informant?"

I explained but didn't go into embarrassing details about the circumstances that brought the deal about.

The First Secretary was irked that a mere contract appointee straight from undergraduate studies in the northern reaches of foggy Scotland had been able to cut an excellent deal with his informant. He appeared envious that I was experiencing first-hand the daily lives of native Romanians and able to function effectively in a repressive communist society. He and his colleagues, on the other hand, possessed all sorts of 'privileged information' supplied by the Foreign Office and the British Security agencies but were not free to socialize informally with Romanians or even to leave the city without written permission. My freedom relative to his and the casual nonchalance with which I was enjoying life in the country, did not endear me to him. Tony Mann, when I told him, including the details, was able to laugh with me.

"Well done, Ronald! Your file says that you were hired for your initiative not your capacity in phonetics!"

Karen kept her word never to enter my life again after the end of my first year. During the beginning of my second, I was walking in the sunshine before classes began, admiring beautiful women on Magheru Boulevard. Suddenly, a fine figure and a familiar face stood out from the passing pedestrians. Involuntarily, I smiled, "Karen!" Karen walked straight past me without as much as a hesitation in her stride. She was a real professional. I sincerely hoped that she had been invited to join the Romanian Communist Party and now had her own apartment.

She did not, however, honour her second promise, i.e. not to make a prejudicial report on me. Karen took revenge for the humiliation I put her through when I rejected her advances. During my second year, the Securitate engaged a homosexual man to pursue me. But how I discovered this, I'll keep for later.

Karen was proud to show me Bucharest and its elegant architecture.

8
BRITAIN'S BIZARRE BOOK PRESENTATION

As part of the bilateral British-Romanian cultural exchange, the British Council donated annually to Bucharest University a package of English language reference books. Tony Mann, asked me to arrange for their formal presentation to the library. The People's Republic of Romania placed enormous bureaucratic obstacles in his path as Cultural Attaché in the British Embassy. He was obliged to make all his contacts through the Protocol Department of the Ministry of Foreign Affairs. This involved a lengthy and capricious process. I, however, as a professor, was not so encumbered.

"I'll be pleased to arrange the book donation," I told him. "But not to the library."

As undergraduates in Aberdeen, the magnificent 15th century Kings College library had offered us a warm place to study and bared our essential reference books in its open stacks. Shortly after my arrival in Bucharest however, I'd inquired where our faculty library was. Professor Ştefanescu Draganeşti had whispered, "In the basement". He made no offer to accompany me, so I descended into the ill-lit depths by myself. I could make out a single tiny square of light at the far end of the corridor coming from the upper half of the single door. I made for the light. On the other side of the open hatch I could see gloomy stacks. I pushed the closed half of the door but it was firmly locked from the inside.

"Buna ziua!" I called. "Good day!"

The head of a dragon-faced lady appeared. "What do you want?" An odd question.

"*Cărți*," I said, "books on semantics." Since Madame Cartianu had switched the course I was to teach, I needed reference books.

"What's the title?" The Dragon-Lady regarded me with suspicion.

"I don't know the exact book I want."

"I can't give you a book unless you tell me the title!"

"What do you have that deals with semantics?" My question was eminently reasonable.

She scowled, disappeared, returned with a drawer full of hand-written filing cards and slammed it down on the counter that divided the upper half from the lower half of the door. A 60-watt bulb cast a feeble glow. The Dragon-Lady positioned herself so that the box of cards was in her shadow.

"Can I search the stacks?" The Dragon-Lady looked appalled. Hostile silence. I'd have to cut to the chase. "Can I speak to the librarian please?"

The Dragon-Lady glared. "I am the librarian! I protect the books from faculty."

"Protect the books from faculty!" There have been memorable occasions on which I've suddenly become aware that my worldview is in direct conflict with that of another. This was one of these occasions. At that moment I understood why the British Council selection board had shown so much interest in my previous experience of other cultures and had selected me and not the best academically qualified candidate. This post demanded extraordinary survival skills. The interview in London had satisfied the board that I possessed such skills.

Politely, I thanked the Dragon-Lady for the valuable lesson in how her library functioned and ascended into the daylight. When I re-entered the common room, I was scrutinised for my reaction to what my colleagues knew I'd been subjected to, but I purposely adopted an enigmatic expression. I learned early in Romania not to allow anyone to read my thoughts or emotions. The lesson stood me in good stead.

The upshot of that experience was my suggestion to Tony Mann that we present the books not to the faculty library but to the Department of English and house them in the empty bookcases that lined the common-room walls. Tony agreed and I arranged the donation, expecting him to make the presentation. "No," he told me, "your doing the honours will consolidate your position in the Department and add to the regard your colleagues have for you."

I reminded him that most of my colleagues would greet me with no more than a "Buna ziua!" when I entered and immediately find an excuse to withdraw. I could also sit at a table in the Faculty Dining Room in the *Casa Universitarilor* and not a single colleague would join me. The country and everybody in it were controlled by fear and my donating a few books could not change that. Nevertheless, he insisted I make the presentation.

All I'd have to do, I thought, would be to speak to the Dean, Ion Preda. But nothing in a communist country is simple and the secret hierarchy within the Department made matters doubly difficult.

"If the books originally intended for the library are to come to the Department of English, the approval of many people will be needed." Ion promised to deal with some of these himself, but I would have to ask for the approval of Professor Tatiana Slama-Cazacu. My belief that Ion held authority within the English Department was false.

"Professor Slama-Cazacu?" I queried. "I've never heard of her and I've been here eight months!"

Ion duly arranged a meeting, but it came with a caution. "They will give you only a moment of their time!"

They? I thought but said nothing. By now I knew better than to ask for explanations.

As we mounted the staircase to her office – I had never heard of a professor having her own private office in the building – we met a man and a woman descending side by side and taking up so much room that others were left with little space to squeeze by. Ion stopped several stairs below them.

"Dr Slama-Cazacu! Dr Cazacu! Good day. What good fortune! Mr. Mackay was just on his way up for his appointment with you."

Both Cazacus acknowledged Ion's greeting but maintained their position above us so that we had to crane our necks like subjects paying homage to a throned Empress and her Consort.

"You may have your say here," pronounced the Consort, "we have important business elsewhere."

We were now blocking the entire staircase creating a bottleneck but the Slama-Cazacus didn't care. Professor Slama-Cazacu's eyes settled briefly on my face and gave me the x-ray treatment. Her Consort was silent. I was clearly not expected to speak. Ion did the little talking that was necessary. Imperial approval must have been given, because we descended the staircase but behind the Empress and her Consort. Ion, looking relieved, whispered, "The book presentation can go ahead!"

When I reached my tutorial room, my students who had all seen me with Professor Preda and the Cazacus, were full of questions. They wanted to know what dealings I had with persons of such elevated rank. I still had no idea who they were and when I asked my students, they were not forthcoming.

Madame Cartianu regretted she was unable to attend the book presentation. She delegated Professor Chițoran as her representative. From my first meeting with him, I judged Chițoran to be a shrewd individual who appeared to be shunned by faculty as much as I was. I learned he was the Communist Party representative in the English Department or perhaps for the entire Faculty. Nobody would talk to me about Chițoran or about how the Cazacus came by their elevated positions.

I'd gathered a little about Chițoran's status by listening carefully. My colleagues never chatted idly. When they spoke, they had thought out precisely what they wanted to say, then they said it, and not a word more. I had to capture the intended information immediately because it would not be repeated. Either I got it or I didn't. Often, I got it, so their respect for me grew.

The day of the presentation arrived. When I entered the faculty room all except Chiţoran were present creating an air of anticipation. Most greeted me from a distance.

Chiţoran made his entrance, dark, confident, self-contained. In a short speech he thanked me and the British Council and abruptly left. Then the moment all were waiting for, the opening of the boxes. One by one, books were unpacked, handed round, and inspected as if they were frankincense and myrrh. After the inspection, two younger women faculty members placed them in the largest cabinet. Then the cabinet door was closed with a satisfying click, everybody smiled. The event was over. I left to avoid embarrassing my colleagues further with my presence.

The following morning, I had to teach at eight o'clock. My routine was to leave my coat in the common-room and then be in my tutorial room 10 minutes before starting class. My students had no reservations about speaking to me there and I looked forward to their company. The informal minutes we spent together, and their uninhibited anecdotes taught me much about how young Romanians lived.

I opened the common-room door, hung up my coat and scarf and turned to admire the new book collection. To my astonishment the bookcase was as bare as the day I'd arrived.

Only much later did I discover that after I'd departed the previous day, the books were befittingly shared out among the faculty members. They would borrow from one another as needed. Leaving scarce and valuable books in an open cupboard was regarded as an act of negligence. Just as I had used my own initiative to subvert the Dragon-Lady's book-control practice, my colleagues had exchanged mine for a practice of their own devising that they knew would work effectively. Romania was full of surprises.

9

SOUR MILK?

Regretfully, I had concluded that *lapte*, Romanian for 'milk' meant '*sour*' milk'. In Scotland, the milk would occasionally go sour, nobody owned a refrigerator in the '40s and we learned to enjoy the flavour. In austere post-War years, no food was discarded. Sour milk was ideal for making pancakes. I'd been living comfortably in Bucharest for some time, drinking sour milk, though foregoing it in my tea, when the kindly Professor Ştefanesu Draganeşti, asked me what I missed most since leaving *England* – the word universally used to mean Great Britain. In most ways, I was better off in Romania than I had been in Scotland, so I had to think very hard.

"What I really miss is fresh milk."

He looked surprised. "You can buy fresh milk every day!"

"I've tried. The milk I buy is always sour."

"Where do you get it?" He asked.

"In the Piaţa Unirii on my way back home around midday."

"It's sold at the door of your apartment building early every morning! Fresh yogurt as well. It's where I get mine."

Milk trucks, he explained, arrived at certain points in every neighbourhood early each morning. At my building, the truck arrived at five-thirty. And so, the next day I rose at five-fifteen and was soon outside my apartment building in the snow. I could hardly believe my eyes! A truck was drawn up on the sidewalk. A mass of people crowded behind it. Romanians preferred a shoving match to an orderly line. I'd seen this at the trolleybus and the tramcar stops. People milled about till the vehicle arrived. When the gates hissed open, they fought for the right to board. I had been given hefty blows by young and old when assuming that it was *my turn* to board. Fortunately, the milk-vendor, a woman whose build suggested that she might moonlight as a wrestler, demanded and successfully imposed a little discipline on her customers.

"Take the exact change with you," Professor Ştefanesu Draganeşti had warned. Apparently, the Wrestler had no patience for making change. I took up a place and let the bodies sweep me forward towards the crates of milk and jars of yogurt were being sold. Meanwhile I was mentally rehearsing my order in Romanian. "One milk and one yogurt please!"

I was close to the truck when an old woman kneed me smartly in the groin and forced me out of line. An older gentleman who'd been watching my relative passivity grabbed my coat, pulled me back into line and held me in front of him. I called my order up to the Wrestler.

She grimaced and said something rapidly that sounded like *"goalie"*. I wondered if rather than a Professional Wrestler she was perhaps a football referee! I failed to understand, made the fatal mistake of pausing, and was immediately shoved aside.

Annoyed that I stood empty-handed, I caressed my injury and watched. "Am I being refused milk because I'm a Westerner?" I wondered. That was unlikely. I'd been subjected to many a curious stare, people had shunned me, but I had never been refused service. Perhaps the successful purchasers had a coupon as well as money. But no, they simply yelled their order just as I had, handed over the exact change and the Wrestler thrust at them what they'd asked for. Annoyed by my failure, I concentrated. I spotted that each client engaged in two simultaneous transactions. As they called out their order, they handed the exact change to the Wrestler and extended empty bottles to her assistant who thrust them noisily into a metal crate.

I lacked empty bottles! That's what the Wrestler had shouted at me: *"Sticlele goale!"* "Empties!"

It has often struck me as odd how, by resolving one problem in an unfamiliar culture, you can be presented immediately with another.

'Now,' I asked myself, *'how can I get empty bottles to exchange for full ones?'* I'd given up buying sour milk at Piața Unirii long ago and had got rid of the empties. So, I waited in the snow until the early-morning crowd had been served and the Wrestler began packing up.

"Vă rog, doamna!" "Madam!" I decided to treat the Wrestler with the utmost courtesy. "Can you let me have one empty milk bottle and one empty yogurt jar?"

She made a Herculean effort to rearrange her scowl into a smile. *"Un leu si douazeci!"* "One lei and 20 cents!" I handed over the coins. Her assistant passed me the empty bottles. Then I repeated my original order: "One milk and one yogurt please!" handing the empty bottles back to her assistant, the correct money to her, and receiving full ones in exchange. I was delighted. Back in my apartment I enjoyed fresh milk for the first time in months and the best plain yogurt I have ever tasted.

Later that morning I told my students about my pre-dawn experience. They thought it was hilarious. I loved them for their interest in the trivia of my daily life. And they, amused by my tales, loved to hear about the simple errors I was constantly making.

That effort of trial and error, observation and detection, was characteristic of hundreds of encounters I had during my first year in Romania. Assumptions based on my experiences in Scotland or in any of the other countries I'd lived in until then – France, Spain, Morocco, the Canary Islands, Portugal, the United States, were of no use. What I had to do was observe carefully, pay attention to the looks and gestures that were exchanged, the words and the tone used. In Romania, patience and observation always paid off.

10

ALL THE BETTER TO KNOW YOU

Soon, I discovered that the few Westerners I met, Embassy staff, the Exchange Professor from the United States, British Council visitors, complained about virtually every aspect of life in Romania. Complaints were founded on their disappointment at seldom having their expectations fulfilled. Their constant complaining made me all the more determined to question my own presuppositions, appreciate the daily challenges Romanians had to cope with, and master what it took to live an enjoyable life in Romania on that country's own terms. To do otherwise was to court doom.

The careful observations I made allowed me to learn how best to conduct myself without offending anyone or drawing attention to the fact I was a Westerner, or causing others distress.

For example, I quickly observed that in casual encounters, Romanians never introduced themselves to me using their full names and never provided any details about their lives such as where they worked, where they lived, whether they had family, or even how they planned to spend their day.

If I bumped into a university colleague in the street, their first reaction was to nod and walk on. If my pointed smile and, "*Buna ziua!*" risked their appearing uncivil, they would take just enough time to shake my hand before hastily going about their business. If a colleague were accompanied by a third party I didn't know, the stranger might act in one of several ways. The most common was simply to disappear. Or they might shake my hand in total silence. Or they might shake my hand and offer their first name only. First names were relatively safe in that I could not identify them to a third party. *"Yesterday I met Ion with Tibi,"* was devoid of useful information without surnames.

Those who offered their first name did so with calculated curiosity in their eyes. Their brain would be working with lightning speed. They seemed to be asking themselves, '*How might this Westerner be of use to me without putting me at risk?*' Then they might ask me a series of questions. "Where do you come from? How long have you been in Romania? How long will you stay? Why are you here? Where do you work?" I would answer their questions without demanding any information in return. The quickest way to bring an encounter to an end was to ask a Romanian a personal question.

The two questions that they invariably ended with were: "How much do you earn?" and "How many rooms do you live in?" Answers to these questions might, I imagined, allow them to calculate how *'important'* I was.

Once I had satisfied their curiosity about me and shown I respected their privacy, they might, very occasionally, express an interest in getting to know me better. It was usually to make a request of some sort, but not always.

Initially, I found the question "How many rooms do you live in?" puzzling but I came to learn that accommodation, or the dearth of it, was on almost everybody's mind in Bucharest. When the Communist Party took over in 1949, there had been war damage and an influx of people from the countryside into the capital. Private property was eliminated and housing was assigned as so many square metres per person. A middle-aged married couple who had reared children in a comfortable family home of and who now lived alone, would be assigned a single room and access to a kitchen and bathroom shared by those families who were allocated the other rooms in what had previously been *"their"* house.

I was invited to one such house as a guest of the former "owners". They had been forced to move into a single bedroom. The other three rooms were occupied by three different couples and all four families shared the kitchen and bathroom. This kind of situation was commonplace. The construction of enormous, characterless, residential blocks around the periphery of the city was one way the Communist Party was trying to resolve the housing shortage.

I invariably had time on my hands during the week and I developed the habit of getting on a tramcar, trolleybus or one of the concertina-buses outside my apartment building and riding it all the way to the terminus. Then I would get off, explore a little and get the next bus all the way back to the city centre. At the end of every bus line I would see enormous blocks of apartments newly finished or still in construction. Families would be moved in even before the builders' debris had been cleared away. I would see men, women and children wading through deep muddy snow from their anonymous block to the bus stop, their galoshes caked in clay.

I undertook these long trips for two reasons. One was that it was a way of seeing more of life in Bucharest than merely walking between my apartment and the Faculty. The other reason was more bizarre.

Diplomats and staff at the British Embassy believed that they were under constant surveillance. Romanians themselves also believed their every movement was monitored. Both diplomats and Romanians told me that I was followed everywhere I went. However, it seemed to me that the task of keeping permanent tabs even on a single individual was probably beyond not only the financial resources of the *Secret Police* but their agents' physical capabilities. Nevertheless, just in case those who warned me were right, I set myself a personal mission to guarantee whoever the *Securitate* agent assigned to my case would be as fully, as uselessly and as frustratingly occupied as possible. My long rides ensured that they had to burn both ends of the candle to keep up with me.

My preferred way of tormenting the Securitate was to make very long trips on the city's public transport at precisely the time I thought an agent might be thinking of winding down his day and heading home to his wife and family for supper. I would mount a trolleybus and gleefully search for the passenger who looked unhappiest or the one who kept glancing despairingly at his or her watch as I rode a lurching vehicle from one end of the city to the other and back.

Another strategy to frustrate the Secret Police had to do with my apartment. I was assured by both Embassy staff and Romanians alike, that my apartment was bugged. I'd quietly open my front door, go out into the hall and ring my own bell. Noisily, I'd welcome one or more imaginary visitors into my sitting room. Once the fictitious guests were seated, I'd engage myself in long conversations, altering my voice to give the impression there were several people present. I'd speak in English, Scots, French, Spanish and my best Romanian. It amused me to think of all the Secret Police these conversations must be keeping busy and how the *Securitate* budget must be suffering. No doubt that they had agents who spoke English, French and Spanish, but how on earth, I wondered, might they translate the Scots?

A Romanian shepherdess spins wool while she grazes her flock.

11

DAILY SHOPPING

Romanians bought the necessities by arbitrarily joining a line or group of excited people. Group excitement suggested but did not guarantee that a shop was open (many shops were closed when they ran out of groceries or goods) and better still, there might be something worth buying. I had learned, like all good Romanians, to keep a string shopping bag in my pocket to be ready for such serendipitous opportunities. Let me explain why.

Shortages

I've been in Bucharest for two or three months. Shortages are common. Without warning any item can disappear off the shelf – bread, shampoo, shoes, toilet-paper. Then, for Romanians, they 'No Longer Exist'! Then, after a week or a month, they make an unexpected return but in limited quantities.

Most shop windows are entirely devoid of goods. I find this only a minor inconvenience. Daily life in Scotland in the '40s and early '50s was punctuated by similar shortages and even when items were available, they were rationed and you needed coupons as well as money to buy them. I did the shopping for my mother from the time I was about five years old and I'd often walk miles from one butcher's shop to another to find anything at all – bones for soup, tripe, a kidney, a strip of liver. In those days, you took what was on offer. I was praised for whatever I managed to bring home.

I remember the day that the last food item, I think it was either butter or sugar, was no longer "rationed". It was in 1954. I was 12 and shopping with our family ration book in my hand. Mrs. Mitchell refused our ration book. "Go home, Ronald, and tell your mother rationing is over!" I did. She was overjoyed.

In 1967, shopping in Bucharest was even worse than the period of rationing in Scotland.

To find out what, if anything, is on sale in any shop I first make sure the shop is actually open. If it is, I join an unruly line just to view the few items on display. I note exactly what I want to buy. It may be a piece of meat, a loaf of bread, whatever, and place my order with a surly shop-assistant. A scribbled order form is thrust at me. Clutching the order form, I join a second line and eventually hand it and my money to a surly cashier. The cashier stamps my order "Paid" and now, with my receipt, I go back to the first assistant and wave the receipt hoping to attract her attention. She takes my receipt and scrutinize it from all angles as if she had never seen one before. If she deigns to recognize the receipt, she wraps the item and thrusts it in the general direction of a third surly assistant who thrusts it carelessly into the crowd. I have to be quick and forceful otherwise somebody else will claim it.

From uncertain start to unpredictable finish, shopping is unpleasant and time consuming. Romanians are far more experienced and ruthless with their use of elbows and knees than I am and so, when I find myself outside again, I'm exhausted, bruised and battered.

Once I arrived home, opened a package that was supposed to contain a quarter-kilo of stewing beef to discover that I'd been given a pig's liver. Fortunately, I am an omnivore.

Occasionally it was unnecessary to even enter the shop. I have seen a truck arrive at a store laden with shoe boxes. Before the driver and his assistant could unload, such a large crowd had flocked around that the shop assistants came outside and sold the shoes directly off the truck. Cleverly, they streamlined the system. One assistant took your money and the other handed you a box with two matching shoes in it. Then, victorious, you withdrew to a safe distance and opened your box. Then you held up your pair of shoes and called out the size you'd been given and the size you wanted. You found a match and made a direct swap. Simple! People seemed to end up satisfied.

On another occasion, I saw a truck surrounded by a hundred struggling shoppers. The shop assistants were selling toilet rolls directly off the tail-gate. Toilet rolls "Had Not Existed" for weeks! I'd learned the strategic use of shoulders and elbows by then and successfully made my purchase. I put my packet of two toilet rolls into my string bag and was stopped a dozen times on my way home.

"Tovarishe!" "Comrade! What've you got there? Toilet rolls? Where did you buy such treasure?" And the delighted comrade, man or woman, would sprint off joyfully in the direction I pointed.

I gave up shopping in stores as soon as I discovered the peasant market in Piața Unirii. For two days of the week that beautiful square became a picturesque farmers' market with all the attendant aromas. Sturdy peasants in traditional dress set up stalls brimming with fresh vegetables, eggs both hens' and ducks' gasping life carp, skinned rabbits, chickens, ducks and turkeys. There were barrels of pickled cucumbers and fermenting cabbage one sniff of which could inebriate you, rolls of sausage and dry salami.

Peasants all seemed to be cut from the same mold. Women were squat, comfortably built, wore colourful headscarves and layers of dark clothes to keep out the cold. The men, taller and slimmer, wore their beautiful karakul căciulă on their heads like tall, rounded tea-cozies. City men also wore the black căciulă but with the crown pushed down.

The căciulă is a beautiful lambskin hat that can vary in colour from glossy black through salt-and-pepper to a light grey streaked with white. The is short with corkscrew whorls that reminded me of aromatic smoke curling from an elegant cigar.

My needs were simple, and I could buy almost all I needed from the market in Piața Unirii.

Country skills were valued in the '60s. Here, the shingle-maker is hard at work.

12

THE BRITISH COMMISSARY

From time to time, members of the British Embassy invited me to make up the corner of a dinner table. Their way of life was constrained by their diplomatic status. Their status constrained their movements and with whom they could come into contact. Travel, whether work-related or for pleasure had to be negotiated through the convoluted channels of the Romanian Ministry of Foreign Affairs. Many Embassy staff simply gave up on travel and lived out their period of service in the confines of Bucharest.

Hence, they tended to take an interest in my daily life because it was so different from theirs. My walks from home to the University, my use of public transport, my local shopping experiences, my local diet, my weekends spent walking in the Carpathians, all that fascinated them. They lived in elegant apartments, rented through Government channels. They had maids and cooks who were almost certainly informants for the Securitate.

The meals I enjoyed in the homes of British diplomats were very different from my own basic soups and stews and seasonal fruit and vegetables. The Western Embassies were supplied with food and other items by a regular deliver truck from Vienna run by a company called Osterman and Peterson. The British Embassy operated a "commissary" in the basement where the goods delivered by Osterman and Peterson were sold to British diplomats and British employees of the Embassy. I was an employee of Bucharest University so did not qualify but since I was able to satisfy all my needs for food locally and was neither a smoker nor a drinker, I had no need of the commissary and thought nothing about it.

However, on one of my occasional visits to Tony Mann in the Embassy, he told me that the Chair of the commissary committee wanted to speak to me.

"The committee," she informed me, "has agreed to allow you to shop in the commissary. You may visit this Saturday morning." She smiled and waited for my jubilant response.

I neither needed nor wanted access to the commissary. However, I appreciated that I was being extended an honour and should express appropriate gratitude. That I would be walking in the Carpathians could not be an excuse for not turning up to accept the distinction that the committee had bestowed upon me.

"Thank you very much," I smiled back.

That Saturday morning, I duly turned up at the Embassy compound and the British guard directed me to the commissary in the basement. Those who were on commissary duty greeted me as if to Cinderella's ball. I was formally warned that I could buy only for my own use; reselling was forbidden. The bad news was that I had to pay in Sterling, a currency I was in desperate shortage of.

Shelves lining the walls were stacked with groceries and dry goods. There were packs of cigarettes and bottles of wine and spirits including whisky. The commissary looked like a corner store in Aberdeen.

While the wives on duty chatted and others arrived to make their weekly purchases, I browsed the shelves. After ten minutes, I approached the exit where two wives were stationed as cashiers. I held a packet of teabags.

"You don't have a basket?" They offered me one. "It's easier than bringing what you want item by item."

"There's nothing more I need..." I started brightly but was stopped short by the incredulous expressions on their faces. There was also the hint of hurt that I might be turning down what they obviously considered to be the greatest gift they could bestow on me – goods from *home*. I immediately felt guilty that I was rejecting one of the perks that gave satisfaction to their constrained lives. Goods from the West were a tiny compensation for the sacrifices they were making in an alien capital behind the Iron Curtain.

In that moment I saw more clearly than I wanted to, the truth of their lives. They were kind and intelligent women, faithfully supporting their husbands, leaving their homes, their families and friends far behind to live in social isolation for two years in an unwelcoming country. They didn't speak the language; they had no access to the legendary countryside; they were not permitted to explore Romania's beauty. I felt truly sorry not only for my own insensitivity but also sadness for the limitations placed on their lives, limitations so brutal that weekly access to simple consumer items took on gigantic proportions.

Adopting my most charming smile, I said, "I'm just joking, ladies!" And to their relief I went back and purchased more tea, some odds and ends, a bottle of whisky and 200 cigarettes. To their delight, I told them that I was tempted to buy more but, at my salary... They clucked their understanding. Relieved that I had not offended, I paid and left. The tea I drank myself, reusing the teabags several times to make them last. The whisky and cigarettes, I traded. Trading, I persuaded myself, was not reselling.

At a British diplomat's dinner table – I was often invited at the last moment – I would hear of food items that were, apparently, unavailable locally and so had to be imported on the Osterman and Peterson truck. Initially, I would offer suggestions where the "unavailable" items could be bought, granted, with some inconvenience. Invariably the diplomats at the table, or their wives, would contradict me. Their maids, they assured me, had looked everywhere but these items could not be found locally. I stopped making suggestions, guessing that their maids found it simpler to have their employers order from Vienna rather than exhaust themselves jostling in crowds or tramping in the winter slush of the peasants' market.

These conversations and observations, however, taught me the importance of always querying second-hand assertions. Assertions are a convenient way of shaping and guiding the beliefs of people who are unwilling to undertake the primary research for themselves. They taught me that a healthy skepticism was my most valuable survival tool. I came to regard every assertion I heard in Romania, whether from Westerners or Romanians, as mere opinion until my own experience proved its reliability.

Communist dictator Nicolae Ceaușescu maintained a network of secret police and informers. In the Romanian Revolution of 1989 he was executed.

13

TAKING STOCK

By the spring of 1968 I felt that the May Day holiday would be good time for my mother, Pearl, to fly out from London where she worked, for a three-week visit. I had two weeks without teaching duties and felt comfortably in control of all aspects of my life in Romania. Aware of my shortage of cash, the British Consul, Doris Cole generously offered me the use of her car as she would be on leave, then.

Judging by feedback from the Dean, Ion Preda, and the generous reactions of my students, my courses were going well. My students were a great source of pleasure. They were young, enthusiastic and highly motivated. When they completed their five years of undergraduate studies, they would be assigned jobs based on their grades. The top positions were in Bucharest with radio and television or with a Ministry. Then came teaching jobs in Bucharest first and then in provincial cities. At the very bottom of the list was school teaching in the smaller villages distant from the capital or any city at all.

Most of my students were from Bucharest and wanted to remain in the capital. They had no inhibitions about telling me of their ambitions during group conversations in my tutorial room. They were seldom alone with me. Because they were young and encouraged by the group, they would tell risqué stories about one another's love lives and laugh uproariously if I blushed. I trained myself not to be embarrassment at the more bawdy tales because if I did, they laughed all the harder. Their love-lives appeared to exceed those of any university student I ever met in Scotland. Or had I lived a particularly sheltered life as an undergraduate? Maybe I'd been blind to what was going on around me, but I didn't think so. Scotland and Romania seemed to be poles apart in matters of passion and affairs of the heart. At nineteen years of age, they talked without embarrassment of their own and one another's "aventuras", their affairs of the heart. However, they were never indelicate nor crude. Their tales suggested romantic intrigue. If, however, I saw one of them in the street, she would launch a charming smile but never stop.

My university colleagues had reached a level of comfort with my minimal presence in the common room. Professor Ştefanescu Draganeşti was one of the very few who conversed with me. He lived with his wife in a street not far from my apartment and we occasionally walked towards the university together. He would walk a few hundred yards with me, then catch the trolleybus. I would continue alone. If we boarded the trolleybus together, I learned not to speak during the journey. To use any language other than Romanian attracted unwanted attention. I had heard Romanians scold speakers of Hungarian or Saxon German. My Romanian wasn't good enough not to have drawn curious eyes and Ştefanescu Draganeşti, like almost all Romanians preferred to attract no attention to himself.

Ştefanescu Draganeşti had taken to asking me to correct the proofs of a textbook he was translating into English. He would occasionally knock on the door of my apartment with a manuscript in his hand. He would apologize, step into my apartment and tell me that the corrected script had to be returned to the State publishing house the following day. "Can you please help me?"

I wondered if he was checking to see who I might have in my apartment but as he was a kindly gentleman whose life had not gone easily when the Communists took over, I felt for him and always helped out. A perk gained from that assistance was my learning how the foreign language courses for school children in Romania were developed. The content was written first in Romanian to the specifications dictated by the Communist Party and then translated into all the other foreign languages with slight modifications permitted to accommodate cultural differences. The English version that I read, contained readings from Charles Dickens' novels presenting an England beset by the inequities of the Industrial Revolution as if they were current in the 20th century. The chapters about life in Romania, projected happy school children spending happy holidays with their happy grandparents on idyllic communal farms or making visits to monuments to the Communist revolution. When I first read this misleading material, I queried its fitness but Ştefanescu Draganeşti became agitated and so I accepted it as blatant political ideology.

I was often invited to dinner parties held by members of the British Embassy and occasionally by the Cultural Attaché at the American Embassy. I never refused invitations from Tony Mann our own Cultural Attaché or Doris Cole the British Consul but I found excuses not to accept many of the others.

My purpose was to maintain distance from the tightly knit diplomatic community to which, by reason of my appointment, I did not belong. Tony had warned me early on that I had no diplomatic status. He had encouraged me to integrate to the extent possible into the Romanian community. Moreover, I found Romania too interesting to limit myself to the Western Embassies. Invitations were usually issued for weekends and I wanted to walk or travel in my free time, so I sent my regrets without causing offence.

When I did accept a dinner invitation, I was quizzed so enviously about the regions I'd visited that I felt genuine sadness at my hosts' constrained lives. They had standing and status, beautiful apartments with cooks and maids, excellent salaries and perquisites, generous home leave, and access to all the food items they were used to back home. What they were denied was exactly what I enjoyed – freedom! Granted, I was the author of my own freedom in that I, on principle, had chosen to ignore the travel restrictions imposed upon me by the Rector's Office. Diplomats had to answer for their behaviour to their Ambassador and to the Protocol Department of the Romanian Ministry of Foreign Affairs; I did not. Their time in Romania was a kind of confinement that denied them insight into the country. They had little direct, first-hand understanding of that beautiful country with its fascinating people from a wide range of rich and varied ethnic backgrounds. It saddened me to discover that what many of these diplomatic families really wanted was be back home.

I enjoyed my regular outings and conversations with Karen. Sometimes Karen, whose resources were considerable, would get tickets for a function – the ballet, the symphony or a film. We would enjoy an evening together among sophisticated and well-educated audiences. Every function I attended whether in Bucharest or in a provincial city was filled to capacity. Educated Romanians – Karen called them the 'intelligentsia' or the 'intellectual class' – appeared to love to dress as well as their wardrobes permitted and enjoy cultural events. Women might wear ancient evening dresses, men threadbare evening suits from more elegant bygone days. Never did Karen introduce me to anybody she knew at any of these events nor exchange more than a *"Good evening!"* with anybody who greeted her. They seemed to sense from her bearing that any words beyond the formal would be inappropriate. So, they would offer a courtly smile, eye me with curiosity, and move on.

I use the word 'courtly' on purpose. The social behaviour of educated Romanians was impeccable. They carried themselves regally, behaved discreetly and conversed intelligently in modulated tones. It was not uncommon to see a gentleman greet a woman by raising her hand to his lips and the woman to respond with graceful acceptance of the favour.

Although it must have existed in Scotland, in my albeit limited and relatively short life, I had never experienced the effortless dignity of bearing, elegance of dress and cultured, informed conversation that I experienced in Romania. Romanians were almost totally devoid of the crass, philistine, uncouth features of Western culture and I admired them enormously for that.

I'd come to know the Prahova Valley and the Buçegi Plateau well, from my walking trips. These trips were usually made on my own but occasionally, when his permit came through, the Vice-Consul, Colin Judd, would join me and together we would cover much more ground than I did alone and explore new routes and wooded valleys.

I'd been befriended by an English engineer, Ron Walker, who visited Romania for a couple of weeks every few months. His job was to inspect the excavators and earth-moving equipment that his company, Hymac, sold to the Romanian Government. He loved the Black Sea coast and we'd drive down there in his car along with his Romanian girl-friend (I was to learn that Ron had "friends" in cities all across Europe, East and West) to spend long weekends in a tiny Lipovan community called Doi Mai. The Lipovans emigrated from Russia in the 18th century. Many were fishermen. Ron was able to rent us tiny rooms in a compound that housed a large extended family who welcomed him with open arms. I was to learn that welcoming arms opened to Ron Walker all over Europe.

And so, with these experiences and levels of comfort under my belt, I began to plan for Pearl, a schedule as interesting and varied as I could.

British Consul Doris Cole's Ford Anglia. She generously lent me her car whenever she went on leave.

14

A LONG WHEELBASE LAND ROVER

One bright day early in 1968, as I was leaving the British Embassy after a meeting with Tony Mann, Bob, one of the British guards beckoned me into his office.

"I've heard you're staying in Romania for another year?" I told Bob I was but that I would spend the greater part of the summer working in the UK to earn and generate some savings.

"Are you thinking about buying a car? Maybe driving back from the UK to Romania?"

"I can't afford that," I told him, wondering if he was hinting that I was stretching the Embassy's patience by so often borrowing the Consul's car and sometimes that of one of the First Secretaries when they returned to the UK on leave.

"How would you like to buy a used Land Rover right here in Bucharest?"

"Sounds great, Bob, but there's no way I can afford it,"

"A British engineer has finished a job here and won't take the Land Rover home with him. I can get it for you cheap."

"How cheap?" I'd never owned a car and didn't really need one.

"£150. It's in good running order."

At £150, I was tempted but I had barely that in my UK bank

account. The British Government reserved its largesse for career officers in the Foreign Office and their minor-league siblings in the British Council. Even overseas students whose studies in Britain were funded by one or other agencies of the British Government enjoyed generous conditions. Contract employees like me, on the other hand, were treated shabbily. My contract within the bilateral cultural exchange agreement required I resign my UK resident status and thereby my social benefits, in return for a miserly £30 paid into my UK bank account. Bucharest University paid me a generous salary but in Romanian Lei which could neither be converted into hard currency nor taken out of the country.

Bob was not to be deterred.

"What if I get it for you for-"

"No!"

"Hold on! I'll get it for you for £100 and find you someone to rent it from you occasionally for Sterling."

"Paid into my bank account in the UK?"

"Paid however you want it."

I liked Bob. He reminded me of my training sergeant in the 3rd Gordons, confident and capable. He had hooked me so firmly that I failed to ask who might rent the Land Rover from me and why.

I arranged to have the £100 transferred from my bank in Scotland to the UK corporate owner of the Land Rover in the UK. Bob called me when the deal had gone through and I went to the Embassy compound to collect my vehicle.

Immediately outside the permanently-open, wrought iron gate to the British Embassy were posted two uniformed Milițieni. They carried rifles. Their principal task appeared to be that of discouraging Romanian citizens from entering the compound and reporting on those who did. Bob and his fellow-guard, both Londoners judging from their accents, controlled everything behind the gates and the wrought-iron fence within which the Embassy stood. The Embassy had once been a private house. The compound included an old carriage house with double doors below and accommodation above.

My experience in the Gordons convinced me that both British had been, or still were, members of Her Majesty's Armed Forces. Their bearing had the self-assured, unhurried confidence of the non-commissioned officers from whom I had received weapons training. They missed nothing.

The Land Rover, parked by the carriage house, was the flat, dull, rust-colour of a low-profile agricultural vehicle, unremarkable and unmemorable. Bob handed me the registration papers. They bore my name.

"That's not my address!" I pointed out. It was *"Something Farm"* in Cornwall, England.

"You don't have a UK address, remember?" Bob explained. "So as not to deprive you of the sale, they agreed to leave the existing address. It's fine." He winked.

He fitted the papers into a metal plate and slid the plate into flanges under the dashboard. The utilitarian interior of the vehicle was no frills metal with canvas seats. "Leave the plate there at all times. If you're stopped, show only your University ID, nothing else."

He handed me the keys. "I have a spare set," he pointed to a row of hooks. Each bore a set of car keys. "Embassy policy!" He winked.

The engine burst into life on the first turn; the vehicle shuddered, rattled and then settled down into a gentle vibration. On the hill farms and moorland estates where I'd worked during the university vacations, I'd been a passenger in Land Rovers. I knew Land Rovers to be reliable vehicles on- or off-road. I drove my Land Rover home and parked it close to my apartment building. Private vehicles were few and far between, so mine sat there alone.

Later that day, my Land Rover refused to start for me. I had neither tools nor mechanical expertise. I did all I could, but it sat in a forlorn spot where I could just see it from my apartment window.

From time to time Bob would call me to say that *certain parties* wanted to rent it. Despite my protestations that it wasn't running, it disappeared for days at a time. After each *"hire"*, a small sum was paid into my account with the Bank of Scotland. I never met those who rented it. The *parties* were satisfied because Bob made no mention of its refusing to start. I was mystified.

When, after a 'hire' the Land Rover would reappear in its accustomed place, I would turn the key. It would run for a few minutes then stop and refuse to fire again. And so, my Land Rover sat close to my apartment building unless rented by *certain parties*.

I took to recording the odometer readings each time it was used. It sometimes covered hundreds of kilometres in just a couple of days as if it had been driven non-stop day and night. Attempting to find out where it might have been, I would take the mileage covered on a given trip, divide by two and draw a circle in pencil on my map of Romania. I poured over the overland portion of the arc which was large and touched on the USSR, Yugoslavia and Bulgaria assuming that it had been driven there and back by much the same route. There was no way to know exactly where it had been, so I gave up speculating and instead, took satisfaction from watching my bank balance grow in small increments.

Secret Exposed

That invisible renters were able to get such excellent service out of my Land Rover but for me it refused to run, seriously bothered me. When he returned to Bucharest in May, Ron Walker triumphantly presented me with a ragged Land Rover manual he'd picked up second hand. Next morning, I dressed in old clothes and tried to start the vehicle with no success. I spent the morning tracing first the ignition system and then the fuel system with the help of the manual.

The ignition system didn't tell me much. The engine turned over but wouldn't fire. The fuel line proved more productive. With great effort, I discovered that welded into the fuel line on my vehicle was a small reservoir led from the fuel tank. A valve had been inserted into the line above the reservoir. The valve could be turned on or off. If turned to the *off* position the fuel in the reservoir would allow the engine to run for a few minutes but then the engine would exhaust the fuel supply and stop. Was this some sort of home-made anti-theft system?

My hands-on experience with diesel engines was limited to a

bulldozer I'd operated one summer at Auch, a sheep farm near the Bridge of Orchy where my cousin Derek worked as a shepherd. The owner, a Mr. Aiken, was an apathetic farmer with few skills and little interest in sheep. His parents had bought the 78,000 acres of Auch in 1940 to avoid his having to serve in WWII. He had lived off government subsidies ever since.

Using the bulldozer, he wanted a small side-stream of the Kinglass River deepened to reduce flooding so that he could graze a small herd of Keil cattle. Diversification, he called it. He hired me to do the job. I'd run out of diesel once and the mechanic, who had taught me how to run the bulldozer, showed me how to bleed the air out of the fuel lines so that diesel could flow again uninterrupted to the injectors.

This idiosyncrasy of diesel engines taught me enough to know that if there was a valve on the fuel line there must also be a manual pump that could be used to clear air out of the line and induce the flow of diesel fuel. I hunted for the pump and found it. Neither the reservoir and the *on/off* valve featured in the manual, so I assumed that these had been added by the contractor to reduce the danger of theft. The device worked because it successfully prevented me from using the vehicle. Bob must have known but for his own reasons hadn't shared what he knew with me.

Turning the valve to open, I pumped fuel through the line until it ran continuously and then hooked the line back into the injector and tightened the nut. I turned the key in the ignition and the engine bust into life. The vehicle vibrated and then settled down to a comfortable purr.

Well pleased with myself, I closed the hood, jumped into the driver's seat and set off up the wide Magheru Boulevard, through the centre of the city and headed towards Herăstrău Park. When I reached the winged statue at Piața Aviatorilor I looked for a fuel station, found one and topped up the tank. I turned and headed back home.

Almost a Revolution?

Driving back down Aviatorilor Boulevard I noticed that it was unusually quiet even for Bucharest. Neither trolleybusses nor trams were running. As I passed Piața Victoriei and turned back on to Magheru, I saw people thronging the pavements. Portable barriers had been set up to hold them in check.

Puzzled, I slowed down to figure out what was going on. These crowds weren't out just to cheer my success. I was puzzled. Suddenly a group of Milițieni, on Motorcycle Police, smart in navy-blue and white, appeared on both my left and my right. An escort! The crowd started cheering!

The Milițieni riders drew in front of my Land Rover and forced me to stop. Masses of people of all ages were cheering and waving flags. Dismounting, the officer in charge approached me.

"Cine sunteți? Ce faceți?" The officer fired questions at me. "Who are you and what do you think you're doing?" I cooperated by showing him my Bucharest University ID. I told him that I was testing out my vehicle after having successfully repaired it by myself. I gave him the modest smile of a mechanical genius. His face turned purple. Angrily, he waved me off the Boulevard. His troop removed a barrier and held back crowds to allow me onto a side-street.

"Park here! Get out! Get into line!" The officer commanded. His troopers all but dragged me out. A small tricolour flag on a stick was thrust into my hands. Now I was part of the crowd.

"Val steagul!" - "Wave your flag!" The officer ordered.

I waved it for all I was worth. I noticed that it wasn't the Romanian tricolour I was used to, blue, yellow and red, but one I didn't immediately recognize, blue white and red.

"There's been a revolution!" I thought, "Ceaușescu's been overthrown. The entire city's celebrating!" I became excited and waved the flag with increased enthusiasm just as thousands of rejoicing Romanians around me were doing.

I was struggling to decide whether I should shout "Long live the revolution"! Or perhaps "Long live the *counter*-revolution!" In Marxist rhetoric, revolutions brought about Communism; toppling a Communist regime would amount to a *counter*-revolution. I didn't want to get my rhetoric mixed up on such a momentous occasion.

Everybody's eyes were focused expectantly in the direction of Băneasa and Otopeni Airports.

"Of course! The new president has just flown in and is staging his triumphal procession to the Parliament Buildings to assume control!"

And sure enough, a vast fleet of immaculate motorcycle Milițieni came into view, ceremoniously leading a shiny black limousine complete with triumphal figure in khaki uniform. "It has to be the new president!" I thought. The flags waved with renewed vigour. Voices cheered. The uniformed figure, a tall, large man with a huge head regally recognised the adulation of the crowd.

I stared more closely. There was something familiar about him. His odd cap with its short horizontal peak. Enormous ears and a huge nose.

Charles de Gaulle!

Of course! The first visit of a Western European leader to Communist Romania. Today! The reason why my classes had been cancelled, so crowds could welcome this old man!

I stopped waving my French tricolour and thrust it at a boy sitting on his father's shoulders.

"Don't forget what almost happened today!" I said with a smile. The boy and his father eagerly accepted my flag. Of course, the ambiguity of what I said was lost on them. General Charles de Gaulle's daring state visit to Bucharest was almost, *almost*, sabotaged. And by an *Anglo-Saxon* to boot! Had I succeeded, it might have been the making of me!

The Milițieni refused to allow me to recover my Land Rover until the following day. As I walked home, I regretted that I had not, after all, witnessed the overthrow of Ceaușescu!

The following day, I returned to discover my Land Rover guarded by two Milițieni. I saluted them, opened the door, climbed in and turned the key. The engine burst into life and I drove to the British Embassy to tell Bob how by dint of persistence and logic I'd overcome the vehicle's problems.

To my disappointment, Bob didn't seem impressed. He was more concerned about the reports he'd had to record into his ledgers when first the *Milițieni and then the Securitate* had lodged a complaint that an unauthorised British vehicle had used the route from the airports to the city on such a historic occasion.

"I'll keep the complaint on my desk," Bob told me. "It needn't go any further. *Sir* would not be happy!" Bob admonished me. *Sir* was how he and Tom referred to the British Ambassador, Sir John Chadwick.

Bob pushed a copy of the daily newspaper towards me. It had photographs of de Gaulle's triumphant procession down thronged boulevards. It also included a line from de Gaulle's address, *"Chez vous, un tel régime est utile; il fait marcher les gens et fait avancer les choses."* / "In this country, your (communist) regime is a useful one – it can get people going and create progress." Alas, the West could overlook much in their pursuit of commercial relations with the communist East.

Two weeks later My Land Rover was rented again. This time, it didn't return.

"Ran into a bit of bother," was all Bob would say.

"My rental fee?" That concerned me most.

"You'll be paid. You'll be reimbursed in full."

And sure enough, I was. Two sums were deposited into my bank account in the UK, the first for the rental, the second of £100 for the value of the vehicle. I was overjoyed and went to tell Bob I'd received the money.

"You never owned a Land Rover in Bucharest, Ron." He winked.

15

PEARL'S ARRIVAL

Sight-seeing in Bucharest

Pearl flew from London's Heathrow to Bucharest's Otopeni Airport and we spent the first week exploring Bucharest, Romania's elegant capital. I chose the most beautiful palaces, churches, houses, galleries, parks and government buildings.

St. Nicholas was originally built as a Russian Orthodox Church at the beginning of the 20th century but after World War II it passed to the Romanian Orthodox Patriarchate under the control of the Romanian Government. The Communist Party systematically purged opposition by the church after it took control in 1947, nevertheless, the Patriarchate placed itself at the service of the government, a pragmatic compromise captured in the phrase I heard repeated constantly, *"A bowed head is not cut off by the sword!"* as if it were proverbial wisdom.

From my own observations in 1967 and 68, Romanians lived daily with 'realpolitik'. Out of necessity, they sought survival using prudence, agility and opportunism – eventually a winning combination.

I counted myself fortunate to admire St. Nicholas Church as I walked along Balçescu and the Magheru Boulevards to my classes. I sometimes entered and sat there as the only worshiper. Approaching the church on a low rise above the main boulevard, Pearl was able to admire the characteristic, multiple, gold 'onion' domes and inside, the beauty of the murals and the iconostasis. So different from our unadorned Scottish churches!

The official newspaper of the Romanian Communist Party was the *Scînteia* – the *Spark*. It had its impressive headquarters in the *"Casa Scînteii", "The House of the Spark"* along with virtually all other publications authorised by the Communist Party. It's an impressive building modelled on Moscow State University. Pearl found the overall effect of its symmetry, grandiose proportions and geometric white stone profile against the blue sky, stark and menacing. She said that the giant statue of Vladimir Lenin that dominated the entrance made her think she was in Russia.

The unusual Mogoșoaia Palace delighted Pearl. The ornate palace was built in the late 17th century by Constantin Brâncoveanu, Prince of Wallachia, in a highly decorative Romanian renaissance style called by proud Romanians, '*Brâncovenesc'*.

The largest and oldest park, the Gradina Cişmigiu offers welcoming paths through flower gardens landscaped around a pond lined by weeping willows. We went there purposely on a Sunday morning when the gardens were at their busiest so that Pearl could appreciate the uses made of the park. Serious stamp-collectors swapped colourful Romanian postage stamps; sober games of chess played between thoughtful men watched by silent onlookers; family picnics enjoyed by the full range of people and dresses that make up this varied nation.

Pearl was intrigued to learn that Romania's multiple ethnic groups reflect the history of a country that emerged, sometimes violently, over centuries from Dacia, a province of the Roman Empire. It has been seen waves of conquerors and settlers from the Ottoman Empire in the east to the Austro-Hungarian Empire in the west and invasions between and during both World Wars. After World War II, Romania was occupied by the Soviet Union. Even after the occupiers withdrew in 1956, Romania continued to be dominated by its aggressive neighbour.

Pearl found the *Muzeul Satului* her favourite spot in Bucharest. This reconstructed "Village" is an open-air ethnographic museum built inside the beautiful Herăstrău Park. The best examples of authentic old buildings were brought from every region of Romania to Bucharest. Peasants' houses; churches; water-driven sawmills, grist-mills and carding-mills; farm houses; inns; even milking parlours, village cheese factories, wineries and plum-brandy stills. All were originally built by village craftsmen out of wood. They represent Romania's rich, traditional, village life.

For the first week, Pearl and I would leave my apartment after breakfast and head off on our explorations of Bucharest on foot or using the excellent public transport system. We would eat a light lunch at a restaurant and then rush back in the late afternoon to shower and dress for one of the many evening celebrations that we'd been invited to as a way of celebrating her visit.

Pearl was in her mid-fifties, in excellent shape and capable of tackling anything. She was a happy, curious, intelligent, well-read Scot who delighted in new experiences especially if they involved interesting people and unusual places. She had the knack of being able to relate easily to anybody, from the meanest farm-labourer to the highest Peer of the Realm. Brought up between the city of Dundee and the village of Coupar Angus in Scotland, Pearl raised and educated my elder sister, my younger brother and me, first in Coupar Angus and then, after 1946, in Dundee. Since leaving our father in 1962 she'd lived in London and worked as a bookkeeper for several substantial businesses.

Pearl was fearless, great company, and open to new experiences. Before going out to one of our invitations in the evening, she would don a wig that suited her perfectly and made her look like a 40-year-old. She invariably won friends with her positive and intelligent conversation, her humour, and her boundless enthusiasm. My mother made me proud.

Doris Cole, the British Consul, invited us to dinner. We enjoyed a quiet evening in her apartment along with two "Queen's Messengers". The Queen's Messengers told us that they were two of only a dozen who deliver, on an as-needed basis, sealed cases containing secret documents from the British Foreign and Commonwealth Office in Whitehall to British Embassies all over the world. Their office as Queen's Messengers guarantees them and their baggage safe passage by virtue of the Vienna Convention. They made Pearl laugh when they told her that the sealed cases they carry are individually identified and each has its own passport! These diplomatic pouches and bags are exempt from all airport checks and cannot be opened, x-rayed, weighed, or investigated in any way by anyone at all until delivered to the Ambassador in person.

When the Queen's Messengers mentioned that they both came from London, the conversation turned to that city of cities. Pearl told them that shortly after arriving in London in the spring of 1962 and getting settled into an apartment in St John's Wood, she had begun attending *Get to Know London* classes every weekend with a professional guide. As a result of these regular walking tours with a knowledgeable guide, coupled with her excellent memory, she impressed us with how intimately a provincial Scot had come to know and love London in barely 8 years.

Conversation turned to the theatre. Pearl and I had seen almost every production on the West End since 1963, surprising both Doris – a keen theatregoer herself – and the two Messengers. I explained that in order to cover the cost of my housing, subsistence, clothing and books while at Aberdeen University, I had worked on building sites in and around London every summer. I'd made enough money not only to save what I needed but also to accompany Pearl to the theatre or on an excursion at weekends.

One weekend we went to South End Pier to be entertained by Stanley Holloway reciting some of his most famous monologues including *'Albert and the Lion'* and *'Sam, Sam, Pick oop thy Musket'*. One of the Queen's Messengers immediately delighted us all by reciting *'Sam's Sturgeon'* That's the monologue where Sam catches a sturgeon in the canal and knowing that the sturgeon is a 'Royal Fish' takes it all the way to Buckingham Palace to offer to the King. The Royal Guard won't let Sam enter the Palace but the King immediately recognises the illustrious Sam at the gate and invites him up to the Royal Suites for a cup of tea. There the King confides to Sam,

> "It's champion seeing thee again,
> But Sam twixt me and thee
> I can't stand Sturgeons
> - But I love a kipper to me tea."

The evening ended in laughter.

As we thanked Doris and bade her farewell, she reminded me that her car keys were with the Embassy guards and we exchanged our respective good wishes for safe and successful journeys – hers back home to the UK and ours up north to the mountains and forests of Bukovina on the southern border of the Soviet Union. We planned to visit the famous painted monasteries built in the 15th and 16th centuries, Voroneț, Suceavița, Humor, Arbore, Moldovița, Dragomirna and Putna.

An iconic example of these monasteries, Voroneț, was built by Stephen III, ruler of the Principality of Moldavia in 1488. He was known as *Ștefan cel Mare*, 'Stephen the Great', for his success in maintaining Moldavia's independence in the face of the expansionist Ottoman Empire. He built Voroneț to commemorate his victory over an Ottoman governor who threatened Christendom. Pope Sixtus IV recognized Stephen as a *Champion of the Christian Faith*.

Voroneț is known as the *'Sistine Chapel of the East'*. The intense blue of the painted frescoes is unique and known as 'Voroneț blue'. These painted churches of Bukovina are listed as World Heritage sites.

Pearl and I were heading off into a very remote and exciting part of the world that few in the West had heard of at the time, an inaccessible and isolated area fought over for centuries by Russians, Hungarians, Poles and Ottoman Turks. We had a car and a couple of weeks, no definite itinerary and an unreliable road map produced by the

Romanian Tourist Board which I was told, at the time, was all that was available.

However, that summer of 1968, while back in London, I purchased large scale, detailed maps of Romania. Many years later these same maps evoked awe in the Security Division of the Royal Canadian Mounted Police who were under the impression that accurate large-scale maps of Romania did not exist!

Pearl Mackay, a thrice-willing travel-companion in Romania. A great traveller, nothing daunted my mother!

16

TRANSYLVANIA AND MOLDAVIA

Pearl and I left Bucharest on a bright spring morning heading for Transylvania via the Prahova Valley. Then we planned to head east to Iaşi in Moldavia and then north to visit the painted monasteries.

The yellow Ford Anglia lent to us by the British Consul was one of the few cars on the road as we crossed the Wallachian Plain towards the Carpathians. We were welcomed by a view of hills blanketed in white plum blossom stretching to the very slopes of the mountains.

Once in the valley and close to the town of Sinaia with its white cottages, we caught our first breath-taking sight of Peleş Castle. Its spires and turrets soar above the forested slopes of the Buçegi range of the Carpathians. The castle was built by King Carol I of Romania in 1873. It became known as the "*cradle of the Romanian nation*" since it was in 1878 that Romania became recognised as a country after it joined forces with Russia to expel the Ottomans. King Carol I's son, King Carol II, was born there in 1893. He married Princess Marie of Edinburgh, a granddaughter of Queen Victoria in 1892. So, as Scots, Pearl and I felt we had a special interest in Peleş. Princess Marie of Edinburgh became Queen Marie of Romania and endeared herself to the Romanian people.

In 1968 there were virtually no foreign tourists in Romania. We had Peleș Castle to ourselves.

We had lunch in one of the villas where I occasionally spent the night after a walking trip. There was no menu. We simply paid a small sum for lunch and a waitress brought the food to our table. Lunch was delicious soup and a ladle-full of bright yellow polenta topped with a wedge of white cheese and butter. We found this as we found all Romanian meals, delicious.

We stopped briefly at Bușteni and Predeal so that Pearl could see the villages that featured so prominently in my weekly letters back home. She shared these with my brother and sister. These villages were starting points for my excursions into the mountains. They were the quiet hamlets where I would alight from the early train from Bucharest, take a deep breath of pine-scented air, and prepare for several hours steady climb up the rocky trails through the forest and onto the Buçegi Plateau. Today, however, our destination was Brașov in Transylvania.

Brașov is an ancient medieval city surrounded by forests and mountains. Parts of its fortified walls still stand. It was built by Saxon colonists invited by Hungarian kings to settle, excavate mines, and cultivate Transylvania between the 12th and 14th centuries. The city has had many names – *Brassovia* – White Water; *Corona* – Crown; *Kronstadt* – Crown City; and in the 1950s *Orașul Stalin* - Stalin City! The streets of the old town are cobbled and the main square is surrounded by elegant baroque-style houses in tasteful shades of ochre, amber and cinnamon. The bells in the tower of its elegant Gothic Church, '*Biserica Neagră',* the Black Church, welcomed us.

I would get to know Brasov and another Saxon city, Sibiu, very well in 1968 when I developed a close friendship with Harald Mesch who also taught at Bucharest University. But that's for a future chapter.

During our two days in Brașov we visited most sites in the city and heard a version of German being spoken by much of the population. Not only did they speak Saxon and look and dress differently, they were all fluent in Romanian as well. Many had a blond, Teutonic air about them. The women wore distinctive blue headdresses and white blouses lightly embroidered with intricate blue designs.

From Brașov we headed north-west to visit the city and the region that had captured my imagination at school, the city of Cluj in the Magyar Autonomous Region. Now, in 1968, the city was called Cluj-Napoca, but people still referred to it as *Cluj*. Despite being the largest city in Romania, it had fewer than 300,000 inhabitants. Like most Romanian cities, it has grown from a medieval, central square.

Preparations for the May Day celebrations were in full swing and the entire city was decorated in bunting and banners proclaiming the beneficence of the Romanian Communist Party. Enormous poster photographs displayed the twisted grimace and broad forehead of Nicolae Ceaușescu the General Secretary of the Party who had taken over from Gheorghiu-Dej in 1965, just three years earlier.

Despite the grandeur of some of the buildings in the centre of the city, their restful ochres and burgundies, the impressive Church of St Michael, and the statues commemorating past national heroes as well as Marx, Lenin and Stalin, we were struck by the constant noise. In all towns and villages in Romania, loudspeakers issued news bulletins and rousing military music from dawn till dusk. We usually became used to it and tuned it out, but in Cluj the blaring music and the strident announcements were so aggressive that we were physically uncomfortable.

Our strategy was to leave the car parked and savour these ancient cities on foot. Since we were the only foreign tourists, both we and the car attracted attention. Nobody would approach us or offer assistance if we looked lost. The car and our clothes warned that we were best avoided.

We parked in the main square and stepped into the cool evening air. A uniformed police officer glanced at the car and his eyes suddenly widened. I knew that he was going to tell us to move the car simply because he didn't want the complication of a Western vehicle with CD plates on his 'patch'.

The last thing that any Romanian official wanted was to accept responsibility for anything at all. To take responsibility risked being blamed if anything untoward happened. Before the police officer had time to order us back into our car and so rid himself of an unwelcome problem, I marched up to him and announced confidently, "Comrade, they told me I must leave my car here, that is correct, isn't it?" The scary, ill-identified *"they"* disconcerted police officers and public servants. It complicated the situation for them and so they usually confirmed my statement so as not to contradict the *"they"*. The officer looked at me and then Pearl and then at the car's diplomatic plates and pointed to a cobbled yard by the side of the church.

"Park there, Comrade!" He said stiffly. "Your car will be looked after."

The passive in Romanian was as useful as the shadowy *"they"*. The person responsible was implied but left unidentified. To all intents and purposes, the real exchange had been:

"The Powers that Be told me I must leave my car here. Do you believe that they were right in telling me to do this?"

"If that's what the Powers that Be told you, then you must park there, Comrade! The Powers that Be will no doubt also protect your car. As you must appreciate, it's got nothing at all to do with me."

I stood to attention and thanked him. Pearl gave him a smile. He walked off, glad to be rid of us.

We enjoyed Cluj but because of the constant blaring from loudspeakers, were happy to leave after lunch the next day. We headed across the plain towards the Eastern Carpathians in the general direction of the city of Iași. There was no need to hurry. We enjoyed the beautiful villages and glorious countryside with fruit trees in full bloom. That evening we spent the night in Peatra Neamț. The following morning, we took our time to explore this quiet medieval town with its tranquil town square. The sunshine and the quiet atmosphere were so relaxing that we remained for lunch.

The mountains were so fresh and green that we were loath to leave them and head east across the plain, so we took a roundabout route to Iași clinging to the mountains as far as Târgu Neamț and only then turning east. Clouds gathered, the temperature fell and it began to rain.

When we arrived in Iași, it was dark, chilly, raining and much later than we'd anticipated. All we wanted was to find a hotel, have dinner and sleep. As chance would have it, we stopped close to the centre right in front of what appeared to be a newly built hotel, a plain, functional, multi-story building with a battery of windows facing the street.

Little did I know it at the time, but later that same summer of '68, as the Soviet Union and all of its Warsaw Pact allies except Romania were poised to invade Czechoslovakia, I would, back in the UK, have occasion to be quizzed about that hotel by an officer of the Royal Air Force responsible for gathering signals intelligence along Romania's north-eastern border with the Soviet Union.

We never booked hotels in advance. Even if our journey had been planned day by day, which it was not, it would have been too complicated to attempt to make reservations. Moreover, I was travelling outside Bucharest in contravention of the warnings I'd been given by the official in the Rector's Office when I arrived in the country. I was adamant that I would give the Romanian authorities no opportunity to prevent me from travelling when and where I wanted by giving them advance warning.

Clumsy, unfettered bureaucracy marred every Romanian service industry to the extent that the client came a very poor second to unnecessary and frequently insane procedures. It was simpler to take our chances on accommodation. So, I pushed open the glass doors and entered the hotel lobby. It was empty and smelled of new construction. We approached the reception desk and the single clerk on duty reluctantly raised her eyes.

"Good evening. I need two single rooms for tonight, please."

The clerk, a middle-aged woman whose appearance would have improved had she smiled, scrutinised me the way that Romanians have perfected. She immediately identified me as a foreigner. For her, the simplest and safest thing to do was to get rid of us by refusing whatever request we might make. A curt "No!" would, as far as she was concerned, conclusively terminate the inconvenience we represented.

"The hotel is closed!" Her words erased me as if I were an unwelcome beggar. She dropped her eyes but not before I had seen the triumph in them.

It was late and I was dispirited by her discourtesy, though rudeness was not unusual in public servants and I was tired by the drive, in bad weather.

I was tempted to find a hotel where we might be offered more courteous service, but I could see that Pearl was tired, hungry and disappointed. I took a deep breath and turned back to the receptionist. "Good evening!" I repeated loudly. She looked up, hostility in her eyes. *How dare a problem, once resolved, repeat itself!*

"Comrade! I told you, the hotel is closed!"

In confident tones worthy of an officer in the Gordon Highlanders I announced: "Madame, this is the hotel, the precise hotel, that my Communist Party Base ordered me to check into when we arrived in Iaşi!" I laid my passport and my University ID card very firmly down on the counter. The phrase "communist base" I knew, would get her attention. Sulkily, she examined my passport and ID without looking at me. She was playing for time, unsure what to do.

"Madame, I can pay the foreign rate for the rooms. Cash in advance! But, alas, I have no local currency." I ostentatiously pulled some American dollars from my wallet making sure she could not see the wad of Romanian Lei I was carrying. Credit cards were not used in Romania in 1968.

As a foreigner, I was frequently charged a rate considerably higher than a Romanian for any service. I was never certain if this was official policy or not, but I didn't find it unfair. Romanians were compensated for low salaries by receiving subsidised services of all kinds. Since I was a foreigner, I contributed nothing towards these subsidies so why should I benefit from them? Besides, since I had a good salary in Lei and had little to spend it on.

I was counting on the receptionist's interest in my American dollars.

"We only accept Romanian national currency." She eyed the dollars in my hand.

"Perhaps, Madame, the hotel might be kind enough to exchange these into Romanian national currency for me?"

I passed her the American dollars. She folded them into her hand

and disappeared for a minute. When she returned she handed me Romanian Lei. She completed the paperwork and I handed her back the Lei in exchange for her typed receipt stating that I had paid for the rooms in Romanian currency. The deal was done! All three of us were satisfied and Pearl, who had followed the odd transaction and knew exactly what was happening, she was an experienced bookkeeper after all. Pearl smiled at the receptionist. To my amazement, the receptionist smiled back.

"Welcome to Iaşi. The restaurant is still open but there is only *ciorbă* left on the menu." Romania is the only country that can compete with Scotland for the quality and variety of its soups. We headed for the restaurant's appealing aromas.

Welcome to Iaşi indeed! By making her day, we had made ours. Romanians were fundamentally a fair-minded people who appreciated win/win transactions. Today we were all winners.

17

PAINTED MONASTERIES

We left Iaşi, drove north to Botoşani, the birthplace of the Romantic poet Mihai Eminescu, and found a hotel. The small town of Botoşani, recorded in 1439, as having been pillaged by the Mongols, no less, would make a central base for our visits to the monasteries. We began early the next morning.

Constructed from the late 1400s to the late 1500s by princes Bukovina's painted churches are famed for the brightly-coloured fresco paintings that cover their external walls. These frescoes are masterpieces of Byzantine art. They were commissioned to educate the faithful in the themes of Orthodox Christianity.

Given our background rooted in the grim history of the Scottish Reformation which gave us unadorned stone churches and austere interiors, we found the painted churches alluring and the rural setting comforting as the tree-clad Scottish glens with their tumbling streams.

The little town of Gura Humorului was celebrating a public holiday and the entire population was out clad in their national dress. Beautifully coloured embroidery covered the breast and back panels of the waistcoats of both the younger men and women. Men wore handsome woollen coats home-spun from natural wool. Their lapels, sleeves, buttonholes and hem were stitched or embroidered with darker wool giving the men a sober, respectable look.

Pearl discreetly drew my attention to one of these older gentlemen. "Ronald! Who is that man the spitting image of?"

"David Sinclair of Abernyte!"

"Right!" Pearl beamed.

David Sinclair, a family friend, had become a successful potato farmer in the Carse of Gowrie on the north shore of the River Tay. He also bred beef cattle and for many years exported quality bulls to estancias run by Argentine-Scots in Patagonia. I asked the gentleman if I could take his photograph. Graciously, he agreed. Some years later I showed David a print and he was astounded at the likeness.

By taking that photograph of the gentleman in the beautiful woollen coat, we had broken the ice with the villagers. Families crowded round eager to show us their beautifully embroidered costumes. Probably because we were in a distant, rural area, people were more relaxed than in the capital. Nevertheless, they asked us no questions in all probability so that they would have little to report to their Party 'base'.

We visited all of the painted churches. Not once did we meet a foreigner from either the East or the West. The few tourists we did come across were Romanian nationals. They nodded but didn't speak to us.

Since we were in the extreme northern part of the country, we wanted to get as close to the border with the Soviet Union as we could, not for any particular reason, but just because it was unknown and different and continued to play such a major role in the world events of our post-war lives! I had a multiple entry-exit visa for the Soviet Union, but Pearl didn't as we hadn't planned to leave the borders of the People's Republic of Romania. We parked the car 100 yards from the crossing point and watched a trickle of traffic held us for inspection. We were thrilled by knowing that what we could see beyond the Humorous armed guards was the Soviet Union, a tightly closed, mysterious and powerful country.

We left Suceavița and reluctantly began to head south. We could have stayed for weeks in the villages of Bukovina, surrounded by mountains and forests, but we wanted to make it to either the city of Gheorgheni or Piatra-Neamț that night. At Câmpulung-Moldovenesc the map showed a shortcut that appeared to avoid a lenghty arc to Vatra Dornei. It meant taking a poorly marked secondary road.

"Let's take the secondary road!" Pearl was always ready for an adventure.

The gravel road started off well, with a thick layer of crushed rock that rattled under the car. As we progressed south, wending our way through valleys and up steep mountainsides, the gravel gave way to dirt with lots of potholes. I was forced to slow down. Driving slowly was no hardship because it gave us time to examine the villages we passed through. Wooden cottages each with its own vegetable garden to one side and flower garden in front invariably with tall colourful hollyhock. Their blooms ranged in colour from black at one extreme to white at the other and all the other colours in between. Since that day, no matter where I am, when I see hollyhock I think of these remote villages in Bukovina, women tending the gardens and men walking home from their work in the forest with felling-axes across their shoulders.

It seemed to Pearl and me that these villages offered the perfect setting for children's fairy tales! Jack the Giant-Killer or Jack and the Beanstalk. Geese stalked the main road of each village, hens scratched, ducks swam on ponds and the head and snout of a pig peered over the fence. A more traditional setting for Simple Simon or Hansel and Gretel couldn't be imagined.

We stopped in the next village to find something to eat but could find not a single restaurant. Nor in the next village. We were not unduly concerned but would have liked to have had something under our belts as it looked as though we would be driving after dark.

All of a sudden, we drove out of the forest and found that the road was reduced to a narrow ledge excavated directly out of the side of the mountain. The drop was at least 300 feet. We could see far over the forest below, beyond the Moldova River into the fast-darkening range of the Obcinele Bucovinei, the foothills of the Carpathians. The road wound up and down the mountainside. The road surface was loose making it necessary to take corners very slowly for fear of drifting out over the ledge. Lights began to appear in the valley some miles ahead and we felt heartened. If there was no hotel, we might find something to eat.

It was dark by the time we reached the village, more a mining camp than a village. It consisted of long dormitory-like structures and a few public buildings including a restaurant with its lights on. Relieved, we parked the car. The aroma of "mititei", heavily spiced minced-lamb grilled, made our mouths water. We'd hit the jackpot!

Inside, the restaurant looked like a Wild West saloon – wooden tables and chairs, a long wooden counter and a huge grill set in an open window. The man attending to the grill looked up in surprise.

"What do you want?"

"Two portions of mititei please!" I gestured to the grill. But now it was our turn to be surprised. Although the large grill sported several hundred skewers of meat rolls sizzling to perfection, the saloon was devoid of customers.

Our second surprise was the man's declaration, "Mititei? They no longer exist!" Pearl and I were astounded. He glanced at us and decided he owed us an explanation.

"Tonight, this camp shows a film in the hall. Everybody is there. The film is due to end now and these mititei are for the families that will rush in here to eat lamb and drink beer for the rest of the evening." He opened his hands and repeated, "Mititei no longer exists! And we have nothing else to offer you either! Nothing exists!"

A restaurant where they had run out of everything! Desperately hungry, Pearl was quick off the mark. She held up one finger and pointed to herself, repeated the gesture and pointed at me. He must have seen the hunger in her eyes because he glanced at his watch, walked over to the grill and picked up two skewers of mititei. He handed these to us along with a chunk of dark rye bread. We accepted greedily. He wouldn't take payment, just gestured for us to leave before, presumably, his act of generosity was witnessed by someone willing to report him for misusing government supplies. I already knew that in restaurants, every item of food on a dish was weighed, recorded and had to be accounted for, so I appreciated all the more this daring gesture of kindness on his part.

As we sat in the car nibbling the delicious grilled lamb, fragrant with herbs, we saw families exit the largest of the halls and crowd into the restaurant, some with sleeping children in their arms. One of these customers might go short this evening but perhaps he would have imbibed enough by then not to notice.

We drove carefully in the dark for the next hour until we joyfully joined the main road and backtracked to Vatra Dornei where we found rooms and enjoyed large bowls of *ciorbă*. To continue south was out of the question at that hour, in the dark and in our state of exhaustion.

Beatifully decorated with colourful biblical scenes, Moldovița Monastery was built in 1532.

18

HOME TO BUCHAREST

We wanted to see as much as we could before returning to Bucharest without re-tracing any part of our route and so we headed in a southerly direction through the villages of Crucea, Brosteni, Borca, Poinana Teiului till we came to Lake Bicaz. Bicaz is a long and beautiful artificial lake formed when the Bistriţa River was dammed to build the Bicaz-Stejaru Hydro Power Plant. We passed enormous quarries and cement plants close to the village of Bistriţa before reaching the city of Gheorgheni where we spent the night.

Gheorgheni, many of whose inhabitants were ethnic Hungarians, is another 14th century medieval city with a colourful history. It had been part of Transylvania within the Kingdom of Hungary and then from 1876 until 1918 part of the Austro-Hungarian Empire. The city was ceded to Romania after the First World War but transferred back to Hungary again between 1940 and 1944. After the Second World War, Gheorgheni became part of Romania once more but between 1952 and 1960, it was included in the Magyar Autonomous Province (that green area on the wall map hanging in my school classroom in Scotland) until ethnic-based sovereignty was abolished by Ceauşescu in 1968 in favour of *Romanianization*.

Both Pearl and I found excitement, glamour, and mystery in

moving from Romanian to Hungarian and back to Romanian culture within the space of a few miles. The people we saw, men, women and children going about their daily lives in these beautiful old cities were clean-cut, formally dressed, dignified, and solemn. We could see the curiosity in their eyes as two Westerners walked their unaccustomed streets. Everybody knew everybody in these towns and even if we hadn't stood out because of the car or my mother's clothes or my shoes (Romanians could spot foreign shoes from a hundred metres) we would have been a focus of curiosity simply because we were not *locals*. Only the enormous portraits of Ceauşescu in the city square and the constant music and rousing speech from the ubiquitous loudspeakers suggested that we were in communist Romania.

The following morning, we continued through the flat ethnic Hungarian region to Miercurea Ciuc in the Olt River valley in eastern Transylvania. Another city with a past almost identical to that of Gheorgheni, Miercurea Ciuc was part of Hungary but the Soviet Army captured the town in 1944. In 1945 it became part of Romania once more. When Pearl and I arrived that day in the spring of 1968 the city was part of the Magyar Autonomous Region, but autonomy was to end that very year and the city would revert again to Romanian control direct from Bucharest.

When Nicolae Ceauşescu came to power in the 1960s, he decided to assimilate ethnic minorities. Assimilation included the confiscation of property and businesses belonging to Hungarians and Saxons and other non-Romanian minorities and the abolition of the Magyar Autonomous Region. Romanians were encouraged to settle in areas that had hitherto been almost entirely Hungarian. The local bureaucracies were staffed by quota proportional not to the ethnic make-up of the region, which would have given the Hungarians a majority, but to the ethnic make-up of the entire country and so favoured Romanians. These measures encouraged Romanians into Hungarian and Saxon areas, but equally effective policies ensured the outward migration of the minorities. Ethnic minority intellectuals were even coerced into leaving their homes and accepting work in predominantly Romanian areas so as to dilute their ethnic concentration even further.

Pearl and I were happiest in the tiny villages and small towns. Because most of the time we were in no hurry and there was little traffic of any kind on the roads to worry us, whenever we saw something of interest, we would simply leave the car and explore on foot. One of these spontaneous stops was made at what looked like an enormous ceramic oven in the front yard of a peasant house in a village in the still predominantly Hungarian *'Ciuc'* region. We parked the car and approached the owner, a middle-aged, countrywoman who appeared about to open the oven. We had visions of loaves of scented bread, roast sucking-pigs, baked geese, or perhaps a haunch of venison. The woman explained that her oven contained no food. She was the local potter and her last batch of pots had been fired and the oven allowed to cool for several days. She had no objection to us watching her open the kiln so long as we kept our distance.

She allowed us to peek inside the oven before she began unloading. We could see plates, bowls and mugs of all shapes and sizes stacked from the hip-level floor up into the kiln's dark, domed interior. As she carefully removed items, we noticed that they all had the same distinctive colours and patterns of blue on ivory. No two dishes were identical. I asked the potter if I could buy one or two. She told me that unfortunately pottery taken immediately from the oven is brittle, "These pieces I am in the process of removing will break before you get them back to Bucharest."

Seeing our disappointment, she went into the house, brought back several flawed pieces that had not been sold from her previous batch, and offered them to us.

"How much?" We asked.

"What do you have to exchange?" We travelled very light with only the barest of essentials and so I was at a loss, but Pearl immediately returned to the car, opened the trunk, and took items out of her shower bag – an unopened bar of soap and a tub of Pond's skin cream. The potter's face lit up in delight. She proclaimed that she now had to give us more ceramics to even the score. Graciously though regretfully, we refused; our space was extremely limited and besides, we already had all we wanted. We left the village potter scratching her head and smiling at the good fortune visited on her that warm spring afternoon with the bees buzzing in the apple blossom of her orchard.

When we arrived back in Bucharest, we were keen to tell somebody about our trip through Transylvania, the Çiuc, Moldavia and Bukovina. Fortunately, we had willing audiences. Two more social invitations awaited us, both from fellow-professors in the University.

The first was for afternoon tea with Professor Ştefanescu Draganeşti and his wife. He'd told me that he lived not far from me, but I didn't know where or indeed anything else at all about his family. My apartment building was one of many that fronted onto a principal boulevard but behind it were streets of decent, single-family villas built shortly after the First World War. Professor and Madame Ştefanescu Draganeşti lived in one of these, or at least in part of one.

They met us at the front door and guided us carefully into pleasant room so full of furniture and bookcases that we had to be guided to a small table already set for tea. In neutral tones, Madame Ştefanescu Draganeşti explained that she and her husband had owned the entire house before the War. In 1944, the Communist Party ousted the pro-German government of Ion Antonescu, and with Soviet assistance, expelled King Michael I. The Communists then took over and ruled as a totalitarian government from 1948 until the present. They confiscated private homes and divided them up into multi-family units. Since 1949, our hosts had been restricted to this single room. They shared the single bathroom and the kitchen with other families who also had rooms in 'their' house. After her short explanation the conversation turned to our trip and how beautiful a country Romania was.

In all the months I'd been in the country this was the first Romanian home I had been invited into, not counting the few uncomfortable moments when Madame Cartianu had questioned my ethnicity, so the details of how Romanian families lived was new to me. I began to appreciate why one of the questions I was inevitably asked was, "How many rooms do you live in?" I now understood the naked envy on the face of my questioners when I told them, "I have an apartment all to myself." They lived in single rooms.

Professor and Madame Ştefanescu Draganeşti were delightful, entertaining hosts. They made Pearl and me feel that we were doing them a favour. They served tea in china cups from a china teapot and offered us tiny pastries on china plates.

Later, Pearl and I agreed that we had been humbled by the graciousness of our hosts, felt profound sadness at their situation, and admired the fortitude with which they bore the years of daily humiliation. I admitted that there had been times when I had shown impatience with Professor Ştefanescu Draganeşti's unannounced appearances at my door and frequent inconvenient demands on my time and felt ashamed for not having shown this kind and cultivated gentleman more consideration.

The poignant memory of that afternoon, the kindness of the couple's gesture to ensure we largely ignorant Westerners felt at home in their single room, and the decades of heart-rending, silent suffering, have never left me.

That same week we were invited to the house of Professor Dan Duţescu and his wife.

To my surprise, Professor Duţescu and his wife lived even closer to my apartment than the Ştefanescu Draganeşti. They lived in a new building only yards from my own, in an apartment that they shared with their adult daughter. I recognised their daughter immediately. She was a junior professor in the English department but had never uttered a word to me beyond "Good Day" before removing herself from my presence as did all her colleagues.

That evening she made up for her previous silence. She admitted that she too was a keen hiker and rock-climber. Though she knew the Carpathians far better than I did, she was delighted to meet a person who shared her enthusiasm for isolated places. After we'd been chatting merrily together with the Duţescus for an hour the doorbell rang. Professor Duţescu left the room. I could hear him greet more than one person. "How will he deal with callers," I wondered, "while two forbidden Westerners are in his house?"

I was surprised when he ushered in Professor Leon Leviţki and his wife. Professor Leviţki was yet another colleague who rarely talked to me. The Duţescus and the Leviţkis were obviously great friends of long-standing. We shared a delightful evening with a great deal of laughter. They were pleased that most of what we had to say about Romania was positive and that we treated the negative with good humour. Of course, truly sensitive matters were avoided by us all.

This was by far the most normal evening Pearl had spent in Bucharest since her arrival and the most normal evening I had spent in the country since arriving many months previously. I realised how much I missed simply talking comfortably with acquaintances.

A day or so later, as we waited in the departure lounge at Otopeni Airport, Pearl raved about her visit, our adventures together in Transylvania, Bukovina and Moldavia, the sights we had seen and the people we had met. We promised to explore another part of this intriguing country together the following year. Tony Mann had indicated that both the British and the Romanian signatories to the cultural exchange agreement expressed the desire that I renew my contract for a second year. I had agreed to do so.

19

YEAR ONE DRAWS TO AN END

The Ultimate Crêpes Suzette!

"Thank you for agreeing to have dinner with me, Mr. Mackay." Mr. Jones of the Foreign Office gave me his best mandarin smile across the elegant table in the vast dining room of the Athénée Palace Hotel in the centre of Bucharest. He had come as head of the British mission to negotiate the continuation of the bilateral cultural exchange between Britain and Romania.

"I understand that you have agreed to extend your service as Visiting Exchange Professor in Phonetics at Bucharest University for a second year, Mr. Mackay?"

"Yes, Sir, I have."

"I invited you to this little tête-à-tête because I want to hear your frank views on your position at the university, your living conditions and your experience in general."

The previous evening I'd dined at Ambassador Sir John Chadwick's residence with the entire delegation and senior diplomats from the British Embassy. As we were leaving, "Mr. Jones" asked if I would have dinner with him in his hotel, the Athénée Palace, the following day. I said I'd be happy to but suggested a different restaurant.

"Where?" He looked doubtful.

"Casa Capşa."

"Why?"

"It has more character."

"Character?"

"It dates from the 19th century. It's very grand. More mysterious, more exciting than the Athénée Palace."

"Mysterious? Exciting?" His voice betrayed alarm.

"Gossip has it that Casa Capşa is the favourite meeting place for secret agents and foreign spies"

He shook his head. "Let's just meet at the Athénée Palace, Mr. Mackay."

And so, here we were, both slightly ill at ease.

It was a slow night for the restaurant, devoid of guests. Although the menu was long and varied, what actually "existed" was limited and neither of us was granted our first nor even our second choice.

As we ate, I told him objectively and without any unnecessary detail that might alarm him, what I enjoyed about the University, my students and colleagues, my apartment and how I spent my spare time.

"You have encountered no challenges? No obstacles?" Whitehall mandarins didn't seem to want enthusiasm; they appeared to want stumbling blocks and pitfalls.

Tony Mann had already reported on certain matters: that my teaching assignment had been altered without consultation and that my apartment and telephone were bugged. No doubt he had reported about Karen, and so I mentioned these matters without embroidering.

"So, you still see this person, this Karen, this informer?"

"She'd come out of this badly if I didn't." I pointed out. "We can attend public events together. It's only every couple of weeks until the end of this year's contract. It's a win-win relationship for us both."

He was silent and inexpressive while he processed this information and then he changed the subject.

"Any restrictions on your movements around the country?" His face was impassive.

I admitted to him that I'd been restricted to Bucharest unless I applied for and received written permission to leave the city. "But I chose to ignore that instruction."

"Chose to?"

"Yes, Sir."

"Why?"

"If I respected that restriction, I'd be doing their job for them, Sir. It's up to them to control my movements, not for me to police myself."

He nodded for me to continue.

"Being stuck in Bucharest would be too restricting, Sir. I didn't want to disadvantage myself by being overly cooperative with the communists."

"So, you are disobeying their rules!"

"They can enforce them if they want to, Sir."

He considered this. "No repercussions to date?"

"None, Sir. I go where I want. I buy a return train ticket at the central station. Sometimes the Consul lends me her car. Occasionally, Colin Judd the Vice-Consul comes with me. Nobody has ever tried to stop me."

He seemed satisfied and changed the subject.

"You appreciate that this bilateral cultural exchange is only incidentally about culture."

"It is?" I had never given the matter much thought.

"Language, literature," he made a dismissive gesture with his hand, "these are merely the soft end, what we expect to follow is commerce, trade, exports."

"I wasn't aware of that, Sir. Commerce isn't my area of expertise." I thought ruefully how recently and at what considerable effort I had mastered phonetics and how that course had been snatched from me by the faded rose that made up my dozen. For the first time I wondered what my 'area of expertise' might turn out to be. I could think of nothing, nothing at all in which I really and truly possessed substantial expertise and the discovery disconcerted me.

He said nothing so I added, "The differences in our political systems must make trade difficult."

"Disagreements about politics, philosophy and economics must not be allowed to stand in the way of commercial exchange." Now he sounded like the chair of the Oxford debating team who had once come to compete with our team at Aberdeen University.

"I suppose not." It was my turn to be doubtful.

"You should be fully aware that our goal is to expand trade with Romania." He paused. "You don't attend "open night" at the Embassy." It was framed as an accusation.

Once a fortnight, on Thursday evenings, there was an open bar in what at one time was carriage house of the Embassy when it had been an elegant private residence. I'd been invited to open-night as soon as I'd arrived and had turned up twice. I barely drank alcohol, was not a fan of ill-lit places and found two or three Embassy support staff and the odd visiting British businessman drinking beer and reminiscing, excruciatingly boring. The fact that drinks were charged in Sterling was also a drawback for me. I was content with my own company, and too interested in Romania to become a regular at the bar.

"Mr. Mann advised me not to become too closely associated with the Embassy." I excused my lack of sociability.

"It wouldn't do any harm to offer your experience occasionally to our visiting businessmen."

I mentioned that I had in fact befriended a British businessman. Ron Walker was the representative of Hymac, a heavy equipment manufacturer in Derby, but far from lending him my experience, it was he who was teaching me great deal. Keen to redeem myself, I told Jones that on several occasions Ron Walker had taken me to the Dobrogea, the low-lying area south of the Danube Delta where Hymac's huge excavators were used to drain vast areas of land and make them fit for agricultural production. What I didn't add was that Ron was proud of the fact that his excavators were being used in place of slave labour provided by political prisoners.

"Mr. Jones" regarded at me sternly. More was clearly expected of me than befriending a single British businessman, so I uttered a vague promise about making a greater effort to attend the open bar more regularly in the future. He let the matter drop.

By this point we'd finished our main course and I was hoping that I could excuse myself and go home. But Mr. Jones asked, "What would you recommend for pudding?" Was he joking, I wondered? But when I looked at him what I saw was a lonely civil servant unaccustomed to being so far from home, a quiet man who liked dessert and so I suggested we order two portions of "clatite". Karen introduced me to "clatite" at our first dinner together.

"Clatite?" Mr. Jones sounded doubtful, so I explained what I thought they were.

"Oh, you mean crêpe Suzette!" He cried out enthusiastically. "A thin pancake smothered with a sauce of caramelized butter and sugar topped with an orange liqueur and served flambé?"

His description of clatite, I conceded, was superior to mine. I had no idea there was a Western equivalent.

With difficulty, I summoned the idle waiter. It was always a challenge to get a waiter's attention even if there were few or no diners. I succeeded.

"Două porții de clatite!" Crêpe suzette for two! I gave the order but was crestfallen that Jones was already familiar with what I'd hoped would be a completely novel experience for him.

The waiter cleared the plates and cutlery from our main course, scraped the breadcrumbs from the tablecloth onto our laps and left. After a very considerable time had passed, he emerged from the kitchen with a large platter containing two pallid pancakes. At his workstation within the dining room, he folded the pancakes, poured alcohol from a crystal decanter over them and struck a match.

There was a *"Whoosh!"* as the alcohol ignited and we watched the waiter triumphantly raise the tray above his head and trot to our table while doing his best to ensure that the billowing yellow flame didn't scorch his hair. I was impressed at the spectacle, Mr. Jones more so.

Good, I thought, at least the flames are brighter in Romania than Jones is used to in England. I made a feeble joke about how the official newspaper of the Romanian Communist Party was called *Scînteia*, the

Spark and that perhaps this flaming offering was the Party's official crêpe Suzette. But Jones hardly heard. His eyes were on the conflagration. Instead of dwindling rapidly to a blue glow and extinguishing itself, the alcoholic flame continued to leap higher!

The waiter, expressionless, extended the tray containing the incendiary crêpe towards us as far as his arms would allow, keeping his head well back. The heat from the hefty shot of burning alcohol was such that the butter and sugar sauce had ignited. The once pale crêpes were fast darkening from ochre to smoky brown and worse. By the time the inferno had extinguished itself, there was nothing left to burn! Two black, elongated cinders lay on the large plate extended towards us. Jones eyes almost popped out of his head.

I knew exactly what was going through the waiter's mind. He had taken delivery of two perfectly good crêpe Suzettes from the chef. At his own service table, he had measured out the alcohol. The startling event had occurred on his watch and there was no question of returning the detritus to the kitchen. What possible solution might he invent to extricate himself absolve himself from all responsibility?

Romanians greatly feared responsibility because it could lead to liability, guilt and then to punishment. The look in the waiter's now startled eyes asked, *"How can I escape responsibility for this act of cremation?"* Inspired, the answer came to him. *"When the Crêpe were in the kitchen, they were the chef's responsibility. When I brought them to my service station, they became mine. But the moment they are placed on the diners' plates they become their property."* And so, with great aplomb and to the utter amazement of Jones, the waiter served one black cinder each, first to him and then to me, bowed graciously, and fled from the dining room.

The event amused me. I also felt rather pleased with myself for having accurately foreseen how the waiter would handle the disaster. Like any self-respecting Romanian waiter in a government-controlled hotel, he had flambéed the crêpe, delighted the diners and, served them to the letter what they had ordered: Crêpe Suzette for two. He was in the clear.

As a boy, I'd eaten many a slice of burnt toast. It had been my job to toast the bread for breakfast. I cut the loaf into slices and watch them brown under the gas grill. Any burned slice was the result of my negligence and therefore nobody but me should have to eat it. My parents and brother and sister had the right to nicely browned slices; black ones fell to me.

Now I was curious to see how Foreign Office Jones would rise to the occasion. Inured by years of my own negligence to the taste and texture of carbonised cellulose, I joyfully crunched my way through the burnt offering on my plate. Jones, the ever-polite, ever respectful English-Public-Schoolboy-become-Public-Servant, followed my example, albeit more slowly and with noticeably less joy. He left more black crumbs on his plate than I did.

Tony and Sheila Mann laughed when I told them how dinner with Mr. Jones had turned out after all, to be a novel experience for him.

A Break in the West

Alexander Dubček was elected First Secretary of the Communist Party of Czechoslovakia in early 1968. Like Romania and the other Eastern European countries, Czechoslovakia had become a satellite country of the Soviet Union after World War II. Dubček began taking steps towards political liberalization, causing the Soviet Union great concern. Now the Soviet Union was demanding its satellites to ready themselves to invade Czechoslovakia should Dubček's reforms threaten Communist Block unity. Romania's Ceaușescu claimed to be unwilling to invade Czechoslovakia.

Romanians seemed to react to this unconventional position taken by their leader in two different ways. Ceaușescu's seemingly independent line encouraged hope but also promoted unease in the minds of Romanians. They lived daily under the fear created and exploited by the *Securitate*, Ceaușescu's network of secret police and informers. Was Ceaușescu's stance a true act of courage or did it disguise a more sinister pact with the Soviet Union? To pursue a more independent foreign policy was Romania being forced to guarantee

Russia that internal dissent would not be tolerated? Nobody appeared to have any clear answers.

I was looking forward to remaining in Romania after the semester ended and to being part of the rising political excitement as the Soviet Union continued to bully Czechoslovakia and the repercussions spilled over into neighbouring countries. However, I had more immediately personal matters to consider.

The most important matter was financial. The British Government deposited only £30 a month into my bank account in the UK and made no contributions towards any social benefits on my behalf. I had saved most of that by living almost entirely on my Romanian salary, but it still meant that I had less than £240 to my name in the whole wide world. Tony Mann was sympathetic to my financial situation and had given me the names of several summer schools in the UK that required instructors to teach English to foreigners. A school in Bournemouth, had offered me £200 a month for a period of 10 weeks. I could if I lived frugally, double my savings before returning to teach my classes in Bucharest in September.

The second factor was that if I remained in Bucharest over the summer, I would be isolated, without even the consolation of the company of my students. Uncertainly over Czechoslovakia was causing Romanians to be even more careful not to compromise themselves with a Westerner. Colleagues in the University encouraged me to spend the summer in one of the seaside resorts on the Black Sea. Their descriptions of resorts flooded with holidaying Swedes and Germans held less attraction for me than gaining financial security. I was considering returning to the UK after my second contract in Romania ended to pursue a graduate degree. Money would be needed to pay for that.

While I was considering my options, a third factor intruded. Ron Walker the Hymac engineer had turned up, unexpectedly as always.

"I'm three days only in Bucharest, Ron, and then I'm driving back to the UK. There's room in the car if you want to come back with me!" That tipped the balance. Economic security, saving the fare back to the

UK, and the opportunity to see more of Europe by car, these reasons convinced me.

"Count me in, Ron. I'm ready to leave when you are!" He was happy to have the company.

We left Bucharest early and headed south-west to Alexandria and then north-west, passing through Rosiorii de Verde, Dragneși-olt, Craiova, Turnu-Severin, and Timişoara.

Ron Walker was a fine companion. He had limitless stories about inspecting Hymac excavators and back-hoes throughout the length and breadth of Central and Eastern Europe. He was always in a good humour and found everything amusing. In addition, he had acute observations to make about Romanians and I was happy to see that many of his analyses corresponded to my own experience. Never glib or facile; he preferred to describe what he saw rather than to be quick to judge. Time and again he had profoundly insightful explanations for events that puzzled me.

I laughed at one of these more light-hearted explanations. Before we left to drive across Europe to the UK, he had invited me on a daytrip to Dobrogea to inspect a fleet of Hymac excavators working on an enormous drainage project in Dobrogea towards the Black Sea.

"Meet me at the Lido Hotel before five in the morning," he'd said. "I must leave at five on the dot whether you're there or not."

I rise early, so I was at the Lido ten minutes early. Ron appeared out of the gloom accompanied by another man, A Romanian engineer who was to accompany us.

The engineer smiled, held out his hand and gave me a cheery, "Chirilă!" I took his hand and gave him an equally cheery "Chirilă!" back, happy to learn a new form of greeting. We reached the project in Dobrogea. The site was a series of drainage canals a kilometre apart. We delivered the engineer to the site office and continued so that Ron could inspect the fleet of back-hoes working at full pace excavating an even deeper channel some kilometres further on.

Now, as we were approaching Arad on the border with Hungary, Ron began to chuckle. "Remember the engineer we drove from

Bucharest to Dobrogea yesterday? He thinks you and he must be related!"

"Related?" How could he think that? Does he have a Scottish background?

"No, but he thinks you share the same name!" Ron continued to chuckle.

"Mackay? Ronald?" I was puzzled.

Ron laughed. "When you met, yesterday morning he introduced himself."

Then it struck me. I thought that I'd learned a new early morning greeting from the engineer: "Chirilă!" and was proud at having returned the same greeting: "Chirilă!" However, the engineer was using his own surname: "Chirilă" and I had replied using his surname: "Chirilă" No wonder the poor fellow was confused and wondered if we had common ancestry!

Ron continued to chuckle about my absurd error until we stopped in Arad.

No sooner had we alighted from the car in an empty square in front of an imposing building that Ron told me was the seat of the municipal government than an excited voice called out.

"Ron!" We both looked up.

There stood was an attractive woman in her 40s, arms open and a huge smile on her face. But it was Ron Walker that her arms beckoned, not me. She unwound herself from the embrace but it was Ron's face that held my attention. He wiggled his mouth and his cheeks a couple of times and his face took on a different appearance. He began talking to the woman in Hungarian. They talked for several minutes, embraced again and off she went with an encouraging swing to her hips. Ron didn't explain the encounter and I used the same rules with him as I did with Romanians: *No questions!*

Ron, I discovered, could speak many languages. Each time he switched, he would adjust his face and take on a different appearance. As a phonetician I appreciated that a particular language demands a particular *set* to the lips, the mouth the cheeks, the tongue, and the teeth. Watch a film without sound. You will see that the facial *set* used by an English speaker differs from that adopted by a speaker of French, or Hungarian.

After a brief rest in the square in Arad, we drove on until we arrived at the Hungarian frontier. Frontiers between Communist countries were heavily guarded. Immigration and customs procedures were time-consuming and exhaustive. Ron had appointments to keep in Switzerland and so all we needed was a transit visa. By the terms of that visa we had to pass through and out of Hungary within 24 hours after being admitted. Nevertheless, the Customs and Immigration insisted that we exchange a large amount of hard currency into Hungarian Forints. The authorities informed us that we could neither convert the Hungarian Forint into hard currency when we left, nor could we take any Forints out of the country. We had either to spend our Forints in Hungary or surrender the unspent portion as we left without compensation. Communist bureaucracy.

In the 1960s there were no freeways or superhighways in the UK let alone in Eastern Europe. We had about 350 miles to cover before reaching the Austrian border. Ron was heading to Switzerland via Graz in Austria by the most direct route and that meant travelling through Hungary cross-country as all of Hungary's roads seemed to lead to the north towards the capital, Budapest. He delegated me the role of navigator because I was an experienced map-reader and we managed to get as far as the city of Szeged by sunset. Szeged, in the great Southern Plain is a distinguished city and Ron decided we would stop for dinner before driving on to Austria to spend the night.

At that hour when the day has unwound and colours take on hues that they're robbed of by the sun, we parked the car by a leafy park by an outdoor restaurant. It was part of a festival because there was also a stage and an orchestra. Couples were seated at beautifully appointed tables. Ron and I sat down.

Service was slow and it was dark by the time the waiter handed us our menus. Most of the diners were eating what looked like an appetising stew with a delicious aroma. Ron asked the waiter what it was. "Birkagulyás!" Came the reply without elaboration. Keen to make time, we asked for two plates of Birkagulyás. The waiter brought them immediately and the scent of the mutton and vegetable stew flavoured with herbs and paprika was mouth-watering. What a glorious warm evening! An orchestra had appeared on-stage and was playing classical music. We were surrounded by well-dressed, sophisticated people in conversation. We had an appetizing meal on the table before us.

A barman was kept busy filling sparkling glasses with purple wine that caught the light. Most of our fellow-diners were drinking wine as they ate. The waiter came over to us with two large glasses.

"Csongrád!" One word. He was not the talkative type.

Ron screwed up his eyes, adjusted his face and began a conversation with the waiter in Hungarian.

"The wine is a local wine from this region – Csongrád," Ron told me. "It complements the Birkagulyás perfectly. Literally, the waiter says that the Birkagulyás *'demands'* a glass of this wine."

"Then we should surrender to the demand!" I was enthralled by the beauty and romance of the evening.

We made fast calculations. If we had a glass of wine, we wouldn't want to drive any further that night. We had approximately 19 hours of our 24 left. If we dined with a glass of wine and then found a hotel and slept the night, we would have about 10 hours left to cover the 300 miles to the frontier with Austria. These 300 miles were not straightforward. They required us to constantly find minor east-west roads linking the predominantly south-north major roads. I might make a mistake and we'd have to back-track. But Ron had confidence in my navigation skills and so the matter was resolved.

"Two more glasses of red wine from Csongrád!" Ron hailed to the waiter.

We had a memorable dinner there in the warm evening at the edge of the park. Hungarian was being quietly spoken all round us, music caressed the darkness. To help rid ourselves of Forints, we tipped the waiter so well that he insisted we had made an error. We shook our heads, thanked him, and left the park. Then we found rooms in a hotel and slept soundly. The following morning, we were breakfasted and on the road by six o'clock heading to Austria. We drove on quiet country roads that passed through villages where white geese, unaware of our deadline, were unwilling to let us disturb their morning walks.

When we reached the frontier on the Austrian border, the immigration and customs police took an inordinate amount of time to inspect and clear us. They had us stand to one side of the car while they removed and examined every item inside with great care. They examined the front and back seats and the trunk. They ran a mirror on wheels under the chassis in case we had anything, or anybody, hidden there.

By the time we reached Graz it was very late. Ron didn't seem in the least bothered by the hour. He drove to a garage on the outskirts, drew in by the unlit gas pumps and switched off the car engine.

"Grab whatever you need for the night. This is where we're staying." He knocked very loudly at the locked door behind the pumps. A window was thrown open and a woman's head appeared.

"Wer ist es? Es ist mitten in der Nacht!" The head complained, extending itself to view the callers.

Ron adjusted his face. "Frau Fritzl, ist es mir, Ron!"

She exploded in a paroxysm of excitement, "Ron, mein Liebling! Wunderbar! Eine Minute!"

The window slammed. The door opened and a delighted woman threw herself into Ron's arms.

I have no idea where Ron spent the night. I slept under the whitest and warmest comforter that I have ever had the good fortune to enjoy. Next morning, the breakfast table was laid as if for lunch, half a dozen different kinds of breads and a score of cold meats and cheeses.

Frau Fritzl gave us both hugs, mine shorter than Ron's. We were off to Switzerland. Whereas Romania and Hungary had virtually no cars on the narrow roads, the wide highways of Austria and then Germany and finally Switzerland were busy with vehicles of all kinds.

Ron booked us into a chalet outside St Gallen in his name and disappeared for two nights leaving the suite for my exclusive use. I spent the days sitting on the wooden balcony overlooking the city with green hills and snowy peaks beyond. Ron returned, paid the bills and we drove to all the way to Antwerp to spend the night. The following day we took the ferry from the Hook of Holland to Harwich. Ron was heading back to his Hymac headquarters in Derby and so we parted ways in Colchester railway station where I could get a train to London. My sister Vivian and brother-in-law John lived in Seven Kings and would, I knew, be delighted, if a little astonished, to see me at their door.

On the train I read Victor Zorza's column in the Guardian about developments in Eastern Europe. Prodded by the Soviet Union, the armed forces of Poland, East Germany and Hungary were massing on the borders they shared with Czechoslovakia. Bulgaria was also cooperating. Romania was not.

Was I only *physically* back in England? My life in Romania seemed so distant that I was beginning to wonder if it had been a figment of my imagination. I looked at the Guardian in my hands and realised that this was the very national newspaper in which I had seen, only a year earlier, the advertisement for the university post I now held in Bucharest. "British Exchange Professor in phonetics." Whether I lectured on phonetics or some other branch of linguistics was merely a quibble. I was the British Exchange Professor in Phonetics at Bucharest University and I would be back in Romania in a couple of months.

Feeling the need to tell somebody, I looked at each of the passengers in my compartment. They studiously avoided eye-contact.

All the experiences I'd had behind the Iron Curtain, I realised, were important only to myself. Bewildered, I watched the telephone posts whip by so close I felt I could have touched them. I imagined myself running up and down the swooping telephone lines, effortlessly matching the speed of the train.

20

SUMMER OF '68

Anglo-Continental

In the temperate, conservative city of Bournemouth I rented a room in a boarding house and taught young mainly Western European adults, five days a week at the Anglo-Continental School of English. The school was housed in a modern building that stood in its own grounds overlooking the town. Teachers had extensive overseas experience in a wide variety of countries though none in Eastern Europe. Some had been earning large amounts of money in Saudi Arabia. Most were teaching for the summer before returning to posts abroad. All had earned post-graduate degrees in the Teaching of English as a Second or Foreign Language, a profession I barely knew existed.

My experience consisted of only my single year teaching semantics and encouraging students' spoken English at Bucharest University, so I listened carefully to my colleagues and began to learn about their profession and the skills it demanded. They told me about the year-long courses they had followed at universities in Leeds or Essex or Aberystwyth. In the evenings, I studied books they lent me.

Soon I was able to figure out that Teaching English as a Foreign Language was an eclectic application of procedures drawn from the broader descriptive fields of phonetics and linguistics combined with assumptions based loosely on educational psychology and pedagogy.

I taught any and all classes that the Principal assigned me and became familiar with how the English language could be broken up into teachable chunks and skills so that students might learn to speak and understand it as well as read and write it in a sort of incremental way. One of the more technologically oriented teachers taught me how to operate the language laboratory and to record exercises on tape suitable for students to practice the sounds, sentence structures, verb tenses and vocabulary of English. I grasped every opportunity to learn.

There was a particular area of expertise that most teachers shied away from, the field of testing students. Someone had to test students' language proficiency on arrival so that they could be placed at the appropriate level, beginner, intermediate or advanced. Later, someone had to test their achievement, that is what they had learned while at the school so that they might be awarded certificates signed by the Principal attesting that they could use English effectively. Students coveted their achievement certificates and used them to impress potential employers.

Most teachers, even very enthusiastic ones, recoiled from these fields of proficiency and achievement testing for the simple reason that they were terrified of working with numbers. Simple arithmetic scared most teachers. They fled in terror when faced with calculating or presenting simple descriptive statistics.

Sensing the opportunity to acquire a competitive advantage, I befriended the individual in charge of student-testing in the school. Will Clouston was a Scot from the Orkney Islands, ten years my senior. He had taken up testing on discovering that the other teachers, let alone the foreign students, were unable to understand his Orcadian accent. When the Principal threatened to fire him, he mastered the skills that others shunned. Now he ran the prestigious "Testing Unit" and had one of the few full-time, permanent jobs available. The school couldn't function without him.

"Become a testing specialist, Ron. Teachers tend to fear numbers, so you'll have little competition," Will advised me. Excellent advice that I took it to heart. It stood me in good stead throughout my career first in applied linguistics and then in programme evaluation.

A second valuable experience for me was the arrival of a small group of Swiss bank employees for a 3-week intensive English course. All were intermediate or advanced speakers of English and became bored with their courses after only a few days. They went to the Principal and asked for more challenging classes 'related to their work'. Perhaps because I was relatively new to the field and had no preconceptions, the Principal called me into his office.

"What do you think we can do for the Swiss students?"

"What do they claim to want?" I asked.

"Something more challenging!" He was vague.

"Why don't I take these Swiss students for the day tomorrow, find out exactly what they want and then, with their cooperation, design a course to suit them?"

"That's not the way we normally work," he warned me. "Students come here so that we can give them what we offer. But your idea has potential and I'm willing to give it a try." It seemed to me perfectly logical that if students were paying, they should get what they wanted not merely what the school had on offer. Instruction should maybe be demand-driven rather than supply driven. From my experience in Romania, the supply-driven shops and the supply-driven economy in general, satisfied nobody.

I had the Swiss students reflect on a single question: "Once you become more proficient in English than you are now, what do you see yourselves being able to do at work that you are unable to do at present?"

Within a day or so each student had a set of individual goals to accomplish that would serve them well on their return to their banks in Switzerland. Where their goals were common, they worked in groups with a teacher to facilitate. Where their goals were unique, I worked with each student separately in the classroom and language laboratory to help them perform the professional tasks they wanted to be able to accomplish.

By the end of the first week the students were delighted. So was the

Principal. He designated me "Course Tutor to Swiss Bankers". In that role, I learned a lot about aspects of instruction that I had never previously imagined. These included learner needs analysis, syllabus design, individualised learning, goal setting, task-based language exercises, materials development, criterion-referenced testing -- in short whatever was needed to teach English as an ancillary skill to professional adults who wanted to be able to practise their professions successfully not only in their mother tongue, but also in English. The Principal invited me back for the summer of 1969 at a higher rate of pay.

The result of these efforts was that I became familiar with identifying the learning needs of students who, themselves, might not be clear as to what they wanted but were smart enough to know that it wasn't what they were getting.

I often went up to London to visit Pearl. She had taken the post of resident domestic bursar in a girls' boarding school in Queens Gate and had an apartment with a guest bedroom. Every second weekend she was off-duty and so we were able to go to the theatre on the Saturday evening and, on the Sunday, explore more of London or go further afield.

The weekends I didn't go up to London, I spent exploring the coast around Bournemouth and Poole. How the tiny sailing boats beat against the wind to inch their way into port fascinated me. I loved the variety of shore birds that fed at the edge of the shore and the mouths of the rivers. Occasionally I went to parties given by teachers who, unlike me with my single room, had rented apartments. From time to time I would invite a female colleague or student out for dinner at an Indian restaurant where I was working my way through the entire menu so as to become familiar with the Indian food I loved. None of my dates kindled any flames although most of my colleagues seemed to be enjoying the liberty that had suddenly hit Britain along with The Beetles, Carnaby Street and the pill.

As soon as a new group of students arrived from wherever, both teachers and students would immediately begin the search for a partner. Students would stay for only a couple of weeks and wanted to pack in as much fun as they could. They wanted to go out with their new-found boyfriends and girlfriends every night, to bars or restaurants on weeknights and to dance-halls, pubs or parties at the weekends. I had neither the desire nor the money to spend several evenings a week out on the town and so I had less to offer a young woman intent on making the best of her brief summer course at the Anglo-Continental.

I was happy to be learning a new profession, spending a couple of weekends a month in London, following Victor Zorza's column in the Guardian and read his blow-by-blow account of the political events evolving in Central and Eastern Europe. I also mulled over what I might do in the future, after my second year's contract in Bucharest ended. And most importantly, I was saving money. On the whole, I was content with my lot.

Statue Dedicated to the Aviators near Bucharest Airport. It was here, the author intruded into the celebrations for Charles de Gaulle's visit to Romania in 1968.

21

SECURITY MATTERS

Strike Command

Before leaving the UK at the end of the summer of 1968 to return to Bucharest, I visited my brother and his family in Hemswell, Lincolnshire. Euan was a Royal Air Force officer, a navigator in Vulcans, the delta-wing strategic bombers capable of carrying out the critical nuclear strike mission in the event of an atomic war. Those who flew the Vulcan were also capable of performing conventional bombing missions. It was a Vulcan crew, who, after the Argentine Military invaded the Falkland Islands in 1982, bombed the airfield at Port Stanley, putting it out of action, and eliminating the Argentine Air Force from the war.

I arrived at Euan's by train. My timing was inconvenient, so he lent me a spritely sports car and suggested I go off and enjoy myself for a couple of days.

'Inconvenience'

With nowhere special to go, I decided to put the sports car through its paces on the A1 and telephone my friend Ron Walker, the Hymac engineer, when I got close to the exit for Derby. I stopped near Newark-on-Trent and made the call from a public booth. Ron at first refused to recognise me. Puzzled, I insisted.

"Ron Mackay! Your Scottish friend from Bucharest!"

"Sorry Ron!" Came his answer. "This is an inconvenient time. We'll meet up in Bucharest in the autumn!"

His reaction surprised me. However, my experience of Ron Walker told me he was a man full of evasions and surprises. Living and working behind the Iron Curtain made you resilient. I put the rejection behind me and consulted my map. I'd go to Oxford, a lovely city with traditions. Travelling in Transylvania and Moldavia had made me realise how much comfort I felt in old cities built of ancient stone. An added advantage was that during university vacation time, inexpensive accommodation would be easy to find. In those impoverished days, parsimony was my watchword.

Terriers!

With the windows open and the car radio tuned to the BBC, I cruised down the A1 enjoying the speed and the breeze. Near Grantham, a soldier in uniform was trying to hitch a lift so I stopped and picked him up. When he asked me where I was going, I said, *"South! I'm making for Oxford."*

With many thousands of miles of hitch-hiking under my belt, now that I was behind the wheel, I felt kindly disposed to those who travelled this way, especially soldiers in uniform. I asked him about his regiment. He was in the Territorials, the Engineers, and was on a special qualifying exercise. He'd left his army camp in Yorkshire that morning and had to get to Brussels in as short a time as possible and using minimum funds. It was a NATO exercise to encourage individual independence and resourcefulness. His promotion, he told me, depended on the outcome.

"I'll take you as far as Watford on the A1." He was delighted.

I told him about my time in the 3rd Battalion Gordon Highlanders, also a territorial unit. We chatted amiably about his training, the weapons he had been trained to use, his current exercise and his ambitions. Because I'd had a similar training, we were able to go into some detail.

"What do you do?" He asked me.

"I'm a professor at Bucharest University," I told him.

He froze. "Romania!" His face was incredulous. "Isn't that behind the Iron Curtain?"

"Yes," I said, secretly satisfied with the mystery and the romance that the phrase 'Behind the Iron Curtain' held for the British. "Romania is on the Black Sea surrounded by other communist countries."

"There's trouble in that region, now, isn't there?"

"Russia and its satellites are poised to invade Czechoslovakia." In the UK we commonly used 'Russia' to refer to the USSR.

He fell silent, faced straight ahead, regarding me out of the corner of his eye.

"Drop me at the next roundabout, please!" We were approaching one of the occasional roundabouts that allowed traffic from minor roads to join the A1.

"I can take you much further south," I reminded him.

"Here's fine," he insisted. As soon as I stopped, he was out of the car in a trice. On reflection, he must have thought that I was a "player" in his NATO exercise, charged with befriending him by posing as a trained infantryman and then telling him I worked in a communist country and was perhaps a fellow-traveller, a communist sympathiser. His "test" would be how quickly he could assess the situation, realise he was being wrung for sensitive information about his mission and take appropriate action.

I regretted that I'd inadvertently and unthinkingly scared that young man and hoped he'd have better luck with his next lift. My two months in the comfort of Britain had taken the edge off my political sensitivity. Slightly annoyed with myself, I decided I'd had my fill of thrills for one day and so I left the A1 at Stamford, and drove leisurely on minor roads through beautiful English countryside towards the south-west and Oxford listening to classical music.

Bletchley Park

Just outside Milton Keynes I noticed that the oil temperature gauge was showing red. Alarmed, I cruised towards a small garage and stopped. The garage owner opened the hood and amid a cloud of steam, found a puncture in the radiator hose. We talked while the engine cooled sufficiently so he could uncap the radiator and replace the hose.

"It sounds like you're from Scotland. On holiday, are you?"

"Yes, I'm from Scotland. I'm on holiday from Romania where I work."

"Ah! So, you must be the short-wave radio operator for the Embassy!" My utter surprise must have shown because he tried to reassure me.

"Don't worry, mate, everybody knows. This is where you guys train, where you get to play with all the latest gadgets." He gestured to a high stone wall and the trees beyond it. I was totally lost for words.

He continued. "Everybody around here knows what those at Bletchley Park do!"

Bent over the engine, he went on to tell me "what everybody round here knows" namely that between the Foreign Office in London and British embassies around the world, coded communications were sent and received by short-wave radio signals. The technicians responsible for the equipment and the coding were trained at Bletchley Park. The technicians, he told me, were also skilled in the detection of hidden microphones and bugs in the telephones and walls of embassies and in diplomats' rented homes.

I only knew Bletchley Park as the centre where, during World War II, code breakers had mastered the secrets of the German *'Enigma'* cipher machine. I knew that there were communications specialists at the British Embassy in Bucharest, that they maintained a 'safe' room in the Chancery and regularly 'swept' the Embassy for bugs.

Shortly after I arrived in Bucharest in August 1967, a senior diplomat in the British Embassy had authorised one of these technicians to visit my apartment. I never met the technician and wasn't allowed to be present when he made the visit. He was able to confirm that my telephone was bugged and that there were eavesdropping devices in the apartment. The news had neither alarmed me nor caused me undue concern since there was nothing that the technician would do and nothing I could do, under the circumstances. Now, here in Milton Keynes, on the edge of Bletchley Park, this village mechanic was calmly telling me far more than I knew about the training of technicians who served the Embassy behind the Iron Curtain where I was working!

Once he'd replaced the hose, I paid and thanked him, then drove slowly past the wall that separated from the outside world, the top-secret modern equivalent of the wartime Government Code and Cypher School in Bletchley Park.

Back to Strike Command

Having enjoyed Oxford, its ivied quadrangles and gleaming spires, I returned to a welcome at Euan's home in Hemswell. I reluctantly surrender his wife's snazzy Ford Capri! That evening he invited me to accompany him to "happy hour' at the Officers" Mess at RAF Waddington.

Euan lent me shirt, tie and jacket because I didn't carry with me the dress clothes to be admitted to an RAF Officers' Mess. Happy Hour was like an embassy cocktail party only chummier but drinks instead of being free, were charged to the officers' accounts.

The Wing Commander and his wife arrived for an aperitif. The WinCo's dutiful wife, spotting a new face, came over and introduced herself to me. I gave her my name and in Romanian style, kissed her hand. She was delighted.

"Where did you learn such courtesy?"

"In Romania."

"Romania!"

"I work in Bucharest."

She looked at me in alarm. "Bucharest? Romania? Behind the Iron Curtain?"

"Yes. Behind the Iron Curtain," I nodded.

Clutching my hand and with the horrified look still on her face, she dragged me to the bar, "Darling! Darling! This nice gentleman has just arrived from Romania! Bucharest! Behind the Iron Curtain!" She emphasised each syllable.

"Yes, Darling. He is Euan's brother. Not to worry. We know about him!"

She regained her composure but continued to give me curious glances.

The following day, Euan took me back to his base at RAF Waddington. I wanted to understand better what he did and if possible, see one of the Vulcans he flew in. By then he'd been in the RAF for 6 years but, because we lived countries apart, enjoyed few holidays and had little surplus income, our paths tended to cross all too infrequently and then only at our mother's apartment in St. John's Wood in London. At the base, he took me along to the QRA room. That's short for Quick Action Alert the exercise where the full complement of aircrew for a Vulcan bomber was battle-ready twenty-four hours a day.

"We can have our aircraft fully prepared and in the air within a few minutes of receiving an alert."

Euan was rightly proud! He introduced me to the crew members, all young, hand-selected, highly trained men like himself. They were dressed in flying kit, reading, or talking. Given the signal, they were ready to run to their waiting delta-wing bomber with its atomic warhead and proceed to their designated target.

Each of Euan's fellow officers introduced himself and talked with me. One engaged me in a longer conversation. After a few minutes I became aware that he and I were alone in the room. The others had quietly withdrawn.

"You've been to Iași recently, I understand?" His tone suggested casual interest.

"Yes," I told him, "My mother and I spent the night in Iași on our way to visit the painted monasteries of Bukovina." I began talking about the painted monasteries, but he steered the conversation back to

The Kilt Behind the Curtain

Iași. I told him we'd arrived in that city late at night, stayed in a hotel, and left for Bukovina the following morning."

But this officer wasn't interested in mere sightseeing. He steered the conversation back to the city of Iași.

"You stayed the night?"

"We did, in a brand-new hotel!"

"Can you describe the hotel?"

I realised that this was not a random conversation about Romanian tourism. This officer had something specific in mind and so I listened very carefully to his questions and what it was in my answers that interested him. Steel reinforced concrete construction and glass interested him. The dimensions of the hotel interested him, its height, its length, its depth. He was interested in its compass orientation; on which sides it had windows, the proportion of concrete wall to glass window.

I described the building, its orientation, all its dimensions including glass and concrete as precisely and as accurately as I could. I had been trained in strategic observation skills in the Gordon Highlanders. The officer seemed pleased with the details I gave him. He thanked me and told me how delightful it was to meet Euan's brother. "Without Euan's navigation skills," he joked, "we'd have lost the Cold War to Russia long ago!" Members of the QRA team drifted back into the room. We all chatted some more, then Euan and I left ahead of a shower of good-natured banter.

It was only decades later that Euan was able to tell me why his colleague had been so intensely interested in that hotel in Iași.

Crews on advanced bombers like the Vulcan, used radar to assist the accurate delivery of warheads to their targets. Radar creates a picture of the ground directly below the aircraft and for a significant distance ahead. The picture helps the crew to identify both their exact position and their target in relation to easily identifiable landmarks such as large buildings and lakes. A concrete wall reflects the radar pulse back to the aircraft; a lake does not. A hotel constructed from concrete will give a strong return but rows of large glass windows on one side will reduce the strength of the return. Ideally, the navigation system needs advance intelligence supplied by "a man on the ground" to predict what the navigator's radar is likely to tell him. The navigator is then able to compare the predicted radar picture with the actual return and so reduce uncertainty about the aircraft's position and guarantee the exact location of his target even if it's a considerable distance away.

Euan's justification for the importance of "human intelligence" provided by the simple man on the ground to complement "imagery intelligence" provided by radar, brought home to me the unromantic nature of the spy. The popular image of the swashbuckling spy in Ian Fleming's dramatic novels is the exception. The work of an agent operating behind the Iron Curtain who discreetly estimates the dimensions of a new building, counts the windows and confirms its precise location, is far less sensational but likely to be equally important than the dramatic exploits of James Bond. Euan's explanation also served to remind me that valuable information can be rendered unwittingly if the questioner is skilled and knows precisely what he wants.

The information elicited from me by that officer in the Quick Reaction Alert room in RAF Waddington would likely have been fed to an intelligence group whose specialty was the creation of "predicted radar maps". These maps would include landmarks previously identified by a human agent on the ground and would help navigators locate their target which, for obvious reasons in the case of Iaşi, they had never had the opportunity to fly sufficiently close to before.

I took leave of Euan the following day. Days later, the communist countries of the Warsaw Pact, Romania excepted, invaded Czechoslovakia, and brought Alexander Dubcek's reforms to a violent end.

22

THE POST-PRAGUE SPRING YEAR IN ROMANIA BEGINS

Return to Bucharest

With some trepidation but great excitement, I flew from London to Bucharest. From the sky, the difference was acute. London was a vast, brightly lit, energetic city. As the BAC One-Eleven circled to land, Bucharest could barely be seen, it was so dimly lit. Through dark boulevards, the taxi took me to my apartment on the corner of Bălçescu and Mărășești. I unpacked my suitcase.

So began my second year in Bucharest. I felt pleased to be back home in Romania and much better prepared than previously, to teach my new students at the University.

Blunder!

That same week, the American Cultural Attaché invited me to a party at the American Embassy to welcome new members of the British Embassy to Bucharest. Relations between the two Embassies was good. Tony and Sheila Mann were still in place. Tony and Sheila Mann, the Consul, Doris Cole, and of course Bob the Embassy guard, were reliable points of reference.

"Ron, I want you to meet two newly-arrived members of the British Embassy!" Sheila Mann presented two couples in their 40s. The men were reserved and clean cut with Thames Valley accents. I was a phonetician after all and took pride in identifying what region of the UK a person was from! They appeared to be slightly overwhelmed at the grandeur of the American Embassy and the many guests.

"Where have you come from?" I asked cordially to get the conversation going, expecting them to tell me the last overseas country they'd served in.

"From England." One man was spokesman for all four. "Our first overseas posting."

Their accents and fish-out-of-water manner told me they were not members of the Diplomatic Corps.

"England! Where?"

The men exchanged glances and the spokesman said, "North of London."

"Doing what?" I was curious.

The spokesman looked uncomfortable. "We were instructors."

"Very far north of London?"

"Not very."

"Near Milton Keynes?"

The spokesman inclined his head.

"Instructors at Bletchley Park? Then you must be the short-wave radio operators!"

The men looked stunned. Tony gripped my elbow and led me out of earshot.

"Ron! What the...! That's not like you! You're usually so diplomatic!"

It had just slipped out. In Britain, my remark would have been laughed off. I would have told the story of how I'd learned, quite by accident, about Bletchley Park and its training role from a simple village car-mechanic who replaced a hose in my radiator. We would all have laughed. But this was Bucharest where nothing was simple, nothing was "laughed off". Life behind the Iron Curtain was serious. At this very moment, the Soviets and their allies were brutalizing their brothers and sisters in Czechoslovakia. There was real concern that Romania might be next. I had blundered.

"I've been gone too long. I'd better switch back into "cautious" mode."

Tony glared at me. It was, I believe, the only misstep I made in Romania.

Good Resolutions

This year, I had made up my mind, I'd do my utmost to make more Romanian friends. I'd try to integrate myself into the daily life of the city and enjoy a broader social life. Given that I was stuck with the stigma of being from the West and that the law discouraged Romanians from associating with foreigners, I realised it was easier for me to make this resolution than to accomplish it. Even if I could, there might be complications, even unfortunate consequences, for the people in question, and for myself.

I'd given this matter some thought. To eliminate complications for Romanians, or at least to minimize the repercussions, I knew I could not be proactive. I'd have to limit myself to a purely responsive role. Romanians, I reasoned, knew the risks inherent in consorting with a Westerner far better than I did. So, if a man or a woman made the first move, he or she had probably already taken precautions to protect themselves from the Secret Police and their network of informants. Of course, it was always possible that they themselves might be part of that system, as the ambitious Karen had been.

My previous year in Romania had taught me that there was much more going on under the surface of people's daily lives than I could imagine. Not having been born and brought up in a repressive totalitarian society under the crushing fist of the Communist Party, I was not equipped to understand easily the intricacies and convolutions of life behind the Iron Curtain.

Accordingly, I reasoned, if I were unable to protect a potential friend, all I had left to protect was myself. Doubtlessly I could, for some misdemeanour real or imagined, be expelled from the country. A greater risk bearing more far-reaching consequences, would result from allowing myself to be manipulated by a woman so that she could leave Romania.

At the Embassy bar, I'd met and talked to the occasional technician come from the UK to install British equipment in a Romanian factory. Two I remember well, were accompanied by their Romanian fiancées. These young men would have left school at 15 and completed a solid apprenticeship training. Perhaps, if they were sufficiently ambitious, they might have completed a Lower or Higher National Certificate by attending evening classes at the local technical college and passing rigorous exams. They would live comfortable and satisfactorily lives in an industrial city in the Midlands and might have married their local sweetheart. Each had, however, ended up engaged to a beautiful Romanian university graduate with years of higher education and bursting with ambition and great expectations from life in the West. Were these relatively naive technicians being used more for a premeditated material end, than for love?

I decided that while I would avoid exposing any friend to danger, my primary responsibility was to myself. This approach, though perhaps apparently self-centred, credited Romanians with the capacity and foresight to look after their own interests, a skill they excelled at. By taking care never to initiate contact with a Romanian I believed I was offering him or her the greatest protection I could. If, on the other hand, they chose to initiate contact with me, I was entitled to assume that they knew exactly what they were doing. If they asked me to adopt additional strategies to protect our relationship, I would. More than that I could not do.

Before the teaching semester began, Dean Ion Preda, as he had done a year previously, invited me to lunch to let me know what my teaching assignment and timetable was to be. He told me it would be essentially the same. I would have first year students, but instead of my classes being spread across four days, they would be concentrated into three. This fitted perfectly into my plans to continue exploring the Carpathians and further afield.

Colleagues at the University welcomed me back but kept their distance as they had the previous year, all except two, that is. I'd met neither of them before and to my joy they made a point, separately, of introducing themselves and conversing with me.

Professor Dino Sandulescu was a scholar some 10 years my senior who had just returned from having spent two years in Leeds University where he had written a dissertation on James Joyce. He was three-quarters Romanian and a quarter Greek. Dino was intellectual, erudite, entertaining and like all Romanians, highly secretive.

Harald Mesch was a serious young man and an ethnic Saxon from Transylvania. I don't know why I hadn't met him the previous year. He taught American literature. Harald was a year or two older than myself and by far the most forthcoming colleague I had met so far.

Both Dino and Harald, quite individually and separately, made it clear that they wanted to befriend me. I had become used to keeping the various strands of my life quite separate. Never did I talk with one person about another or even mention the name of a third party. Never did I reveal or pass on information no matter how innocuous it might appear. I simply kept the different people in my life and the different parts of my life isolated, entirely sealed in separate boxes and never, ever allowed the contents of one box to percolate into the other. Romanians were hypersensitive to anything that might put them at risk. They both recognised my caution, appreciated it, and as a result were willing to place a little trust in me. Of course, I had no way of knowing if they were informants or not. I trusted that they were not.

23

AMERICAN COUNTERPARTS

The visiting American professors of literature I met, Dale in '67 and now Tom Fitzsimons in '68 seemed to have difficulty acting in a way that put their Romanian colleagues at ease. Tom Fitzsimons was the worse. His blatant failure to understand the need for a minimum of discretion could cause a Romanian colleague to freeze and cringe inwardly, but he was oblivious to the impression he made.

One day, in the faculty common room, Tom, in a loud voice, invited Ion Preda and me to the coffee shop on Magheru often used by faculty members. To my surprise, Ion accepted. Once in the coffee shop, Tom began to talk in his leisurely drawl hushing all other conversations around us.

"Well, I dunno, Ion, you say that things are like this in Romania but let me tell ya, just yesterday, I was talkin' to Dan. You know Dan? Dan Duțescu? 'Course you know him! An' Dan, he was tellin' me the very opposite. And Leon too. You know, Leon Levițki. I was talkin' to him and I think he would disagree with you too. Now what I wanna know is this, how is it that two or three people, Romanians, teachin' in the same department, can tell me two completely different things, eh? How, Ion? Go on, explain that to me! I really, really wanna know, Ion. I'm serious! I wanna understand! I wanna know what makes you communists tick!"

Ion visibly shuddered, looked at his watch, politely excused himself and fled!

Tom, puzzled, shook his head. "Shit Ron! Did I say sumpin? Do you get these guys? I don't get these guys! I don't! Gotta be sumpin', sumpin' in their diet, maybe that çiorba or the polenta stuff. Maybe it's just the commie way of life they got here!" He shook his head in despair.

And I thought to myself, "You bet, Tom. An' you just ain't never ever gonna get 'em!"

That is why, or at least one reason why, without being uncivil, I preferred to keep my distance from Dale and Tom or meet them one-on-one. On their own turf, Americans like Tom were outgoing, generous and delightful.

The other reason I kept my distance from Tom Fitzsimons was because he seemed to consider himself quite the ladies' man. He was married, but not accompanied by his wife. She may have been happy to be free of him for a year. Tom tended to force his attentions on the younger female faculty members.

Women faculty members, at least those who used the faculty common room appeared to enjoy a lesser status than the men. Status was signalled in subtle ways. One was the area of the common room they made use of. The senior, more prestigious male faculty members claimed the area closest to the door; the female professors frequented the more distant part.

When I entered, I would hang up my coat, stand close to the door, greet those already in the room and leave. The women would never come forward to greet me and I took that as a warning that I should respect their privacy. Tom liked to wade through the room and latch on to the younger women. He would make a beeline for one in particular. She gave the distinct impression that his attentions were not particularly welcome. He was unable, or unwilling, to recognize the signals she was giving him.

Tom knew I could have the occasional use of a car from the

Embassy and he pestered me to take him and this young woman on a weekend trip to Braşov.

"How do you know she'll accept?" I asked him.

"I just know these things!" He liked to display his worldliness.

Tom's advances must have been rebuffed because I ended up taking only him to Braşov and spent the weekend feeling like I was babysitting a particularly demanding, undisciplined child.

However, I was only a month into my new semester's teaching when one of my students, the most exotic, glamorous and alluring young woman in the class, made an unexpected move.

The author spent many happy days in the challenging Carpathian Mountains above the Prahova Valley.

24

REGARDING ROMANCE

Doina

It was autumn 1968. I had been teaching my new group of first-year students for more than a month. Walking to class at seven-thirty on this dreary day, I was reminded of the murky mornings of Old Aberdeen despite that city being on the 57th parallel and Bucharest being further south on the 44th, the same latitude as Genoa and Bordeaux. I arrived at my tutorial room fifteen minutes early.

This year's students were as delightful as my previous year's had been. We bonded well and were working hard as well as enjoying the classes. Most were young women but there were a couple of young men and one older man of about 30. All were bright and enthusiastic, the women even more so than the men. The exception was the older male student who behaved respectfully but never asked questions and appeared politely bored. The other students tended to avoid him. I could understand that at ten years older than the average he probably found his classmates immature, so I tried hard to relate to him and encourage his active participation. He hung around me more than I wanted, especially during the break when I preferred either to be on my own, or with my student group and enjoy their banter.

As I entered my tutorial room that morning, it was dismal. To brighten things, I went to switch on the lights. As I reached for the switch, a woman's voice came from the gloom.

"Professor Mackay, please don't open the lights."

Startled, I paused. Doina, the most exotic, the most glamorous, the most alluring student in the class was half-hidden in the shadows.

"I was just thinking of kissing you!" She, bold!

"Not a good idea!" Me, startled!

Truth be told, I was intimidated by Doina's very apparent physical charms. During class, she had the habit of subtly moving her body so that her outstanding attributes were impossible to ignore. I sometimes wondered what a girl like Doina was doing in an English programme. In London or Paris, she would have been securely engaged to a successful barrister, or to a medical doctor at the very least, and would never give someone as ordinary as me, a second look.

"Then where?"

Stunned by her temerity, I had no answer.

"Invite me!" Teasing.

And so, to my utter astonishment, I did invite her. We dined in Capșa's the following week. Every covetous male eye, and many a jealous female one, was on Doina from the moment we entered. Now I really knew what it was like to be James Bond! She flirted delightfully with me. I was overwhelmed and flattered even though I knew she was flirting not with "me" but with some chimera of her own imagining.

After an hour and a half, to my consternation, our conversation began to dry up. I was becoming bored and I suspected Doina was feeling the same. It was a weekday; we both had classes the following morning. I mentioned the hour and then asked the waiter for the bill. Doina leaned intimately across the table, "You will walk me home?"

She led me to quiet streets off the boulevard and stopped in front of a row of once very fine mansions.

"You live here?" I was unfamiliar with such grandeur.

"Close by. But you must leave me here."

I leant towards her for the kiss I'd coveted all evening.

She drew back, "Not here! But we will! Next time! You will take me to your apartment!"

What a promise! I watched her haunches swing seductively, towards her invisible home.

As I walked home, I asked myself questions. "Do you really want an affair with someone who, under normal circumstances, would never give you a second glance?" "What it would be like to have an affair where the only chemistry is physical attraction."

"No!" and "Pointless!" I had little difficulty answering my own questions.

I never again invited Doina and Doina never again lay in wait for me. Like me, she may have found the answers to her own questions. Perhaps she too gave thanks to have avoided a pointless relationship.

Six months before the academic year ended, I learned that Doina had found herself a designer-suited Italian businessman many years her senior, who was expanding his enterprise to Bucharest. She was seen, I was told, driving his bright red sports car around Bucharest and waiting impatiently for permission to leave Romania and marry her Romeo.

"Good for you, Doina!" I thought. And after a moment added, "Good for you too, Ronald!"

Nevertheless, I envied, just a tiny bit, the Italian entrepreneur his adventure with the sumptuous Doina.

Serious Students

All of my students at the *University* were serious, hard-working and remarkably proficient in English despite never having spoken to a native speaker in their lives before attending my class. Several, all of them women were outstanding. They wanted to perform not just well, but perfectly and were willing to put in the effort required.

Having spent two summer months in Bournemouth teaching English to Western Europeans, I'd been able to compare the characteristics of different nationalities. Many of my Bournemouth students were enthusiastic but few set themselves really high standards and fewer strove to reach them. The separating factor seemed to be national culture. Spanish students for example brimmed with enthusiasm but generally appeared content to reach a very mediocre level of proficiency and then stagnate. German students on the other hand were less ebullient but strove for ever-greater accuracy and precision.

These differences were evident as soon as students at the Anglo-Continental School took the proficiency test that allowed the principal to assign each to a class appropriate to their proficiency level.

A Spanish student, believing that he had been assigned to a level lower than that which corresponded his self-image, might complain vociferously:

"Meester Mackay! Is beeg meestake! I eena helementary, butta helementary eesa mucha too easy fora me! I wanna hintermediate! You esee how good I espika Inglis?"

On the other hand, I remember a German student, representative of her nation, whose proficiency score had placed her in my advanced class. At the end of the first day, she waited until the other students had left, approached me politely and at a comfortable distance, expressed her concern calmly and clearly.

"Mr. Mackay, I am sorry. I think there has been a mistake. I have been placed in the advanced instead of at the intermediate level. Might you arrange to accommodate me to an intermediate class? I have so much to learn."

Romanians liked to think of themselves as Latin. I knew the Latin people of Spain and Portugal well. Romanians spoke a Romance language, the only Latin-based language that evolved east of Rome. They demonstrated the enthusiasm of the Spanish. However, Romanians, in my estimation, also possessed the discipline, industriousness, self-control, conscientiousness and persistence of the more northerly nations. They had a keen capacity for objective self-evaluation and a powerful drive for self-improvement.

Every time I heard Romanians say they were like the Spanish or the Italians, I believed they were selling themselves short. They had the educational and cultural standards of the very best of the West. Most of my students could have competed favourably with their counterparts in any British university.

Superlative Students

The three most outstanding students in my class were distinct in looks personality and character.

Astrid was a modest, dark-haired young woman, who undertook her tasks very quietly and produced near-perfect work. 'M' was Astrid's antithesis; she conducted herself in a self-assured manner, had an open, smiling face and clear bright eyes. 'M' showed charismatic leadership, and constantly surprised and impressed me with her quick mind and sharp intelligence. The third, Liliana was totally unlike either Astrid or 'M', indeed she was unlike all other students. She cared little for her appearance, had a mind like a steel trap, an impressive memory, an enormous capacity for learning and was teaching herself Japanese so that she could be hired by the Government to host visiting Japanese businessman. Invariably she produced brilliant essays that were double the required length. However, she possessed few interpersonal skills.

Most of my students were not yet 20 years old and radiated the enthusiasm and freshness of youth. All appeared to get on well together and showed amused tolerance for one another's foibles, including Liliana's eccentricities. There appeared to be little or no jealousy and they were constantly making fun of one another. Whoever happened to be the butt of the joke never took offence but cheerfully joined in the fun even if the laughter was at their expense.

25

THE LIPOVENI AND MURFATLAR

Most weekends, I went walking in the Carpathians but when Ron Walker occasionally visited Bucharest, I would adjust my plans. Ron, his Romanian girlfriend and I, would drive in his car to the little Lipoveni community of Două Mai on the Black Sea coast. It was Ron's favourite spot. On one of these trips, Ron invited Dave, a Scottish engineer who was overseeing the installation of a food-processing plant in the north of the country. He spent alternate months in the UK and at the factory site.

I had never met Dave before. He was in his late 50s, a very practical, very fit Scot from Glasgow who had left school at 15 and completed his apprenticeship as a millwright with a British business that manufactured mechanised systems for processing and canning food. As the company grew and began to export its equipment, Dave became their chief trouble-shooter. He had never married. He had a longstanding, very pleasant, Romanian girlfriend. Anca was a divorcée in her 40s who lived in Bucharest and with whom he spent weekends and holidays.

The five of us would leave Bucharest late in the afternoon on a Thursday and reach Două Mai late the same night. They all enjoyed Murfatlar wine that too often *"no longer existed"* when they asked for it in Bucharest. The wine was made close to the Black Sea coast. One day when the five of us were walking through Două Mai together, we passed an unusually attractive shop. Most shop windows had nothing on display but this one looked like an Italian delicatessen filled with crusty bread, great balls of yellow cheeses and – surprise! – bottles of Murfatlar wine!

The girls were ecstatic! "Look at the price! That's a fraction of what we pay in Bucharest!"

"That can't be right!" Dave was the pragmatist.

"It's right!" Eva and Anca insisted. "It's inexpensive because Murfatlar is produced in this region!"

"Let's ask," Ron suggested. And so, we did.

"That Murfatlar in the window, is it really just 12 Lei a bottle?"

"Twelve Lei the bottle," the shopkeeper confirmed.

Ron and Dave were for buying a bottle each along with some bread and cheese and holding a lunch on the beach. The girls, overawed by the low price, insisted they buy a full case each. So, Ron and Dave bought a full case each, stowed these, the bread and the cheese and some glasses bought in the same store into the trunk and we drove to the beach.

We spread our coats on the sand and sat down to enjoy lunch despite the inclement weather. The girls handed round bread and cheese. The wind gusted and the cheese became gritty.

"Hurry up and open that wine, Dave! We'll need it to wash down the sand!"

Dave finally got the cork out of one of the bottles, poured, and glasses were handed round.

"Cheers!" Each of us took a large gulp to rinse the grit down. There was an explosion of coughing and spluttering.

"Oțet!" "Vinegar! Yuk!"

"This wine is off!" All agreed.

"Open another bottle!" Dave did. Poured.

"Cheers!" The coughing and spluttering were repeated.

"See! I knew there had to be a catch. The price was too good to be true!" Dave laughed.

Eva and Anca were affronted. "No!" They insisted. "No catch! The wine's off! We'll take it back. The shopkeeper must replace it with good wine!"

Ron re-corked both bottles, walked up the beach to the car and drove off back to the delicatessen with Eva and Anca. Dave and I sat on our coats with our backs to the rising wind and spat grains of sand.

When they returned, Ron was triumphant, the girls sheepish. Each carried a bottle of water. They had thumped the boxes of wine down on his wooden counter and complained. Indignant, the shopkeeper had pointed to the label over which a purple stamp announced that the contents were 'Vinegarised Murfatlar'.

"The shopkeeper accused us of being illiterate!" The girls were affronted.

Our lunch consisted of bread and water gritted with sand. None of us was satisfied.

We deposited the 'Murfatlar' at a convenient spot on our way back to Bucharest.

Dudu with Lipoveni friends in Două Mai, Dobrogea on the Black Sea Coast.

26

FOUNDING FRIENDSHIPS

"Operation Danube", the so-called Warsaw Pact invasion of Czechoslovakia in August of '68 was, as far as I was able to gather, a Soviet invasion. The small military divisions from Poland, East Germany, Hungary, and Bulgaria were integrated into the Soviet Army commanded by a general of the Soviet Armed Forces. The USSR was taking no chances!

When I returned to Bucharest shortly after the invasion, I could feel the subtle changes that the event had brought about in Romania. There was a rising level of uncertainty among Romanians about their immediate future. Ceaușescu's government had refused to allow the Romanian army to participate and even denounced the invasion as "aggression". For some time, Romania had, for trading purposes, been cozying up to traditional enemies of the USSR, China and Israel. Israel had captured many Soviet-built tanks when it invaded Egypt in 1967. It was rumoured that Romania had been supplying Israel with the necessary spare parts to repair these tanks. Would the USSR use Romania's behaviour as the excuse for a military invasion from the north? In such an event, Romania would receive no help from the West. Romania was surrounded by only Warsaw Pact countries that had already collaborated to invade Czechoslovakia. In the eyes of the USSR, Romania might be nothing more than an obstinate and unruly neighbour. To preserve its own existence, the Soviet Communist dictatorship had already shown it was prepared to stamp out any dissidence within its own territory as well as in Central and Eastern Europe.

After my return to Bucharest, I continued to follow events by reading Victor Zorza's daily column in the Guardian at the British Embassy. I also read the local Romanian press and watched the news on the television set in my apartment. No clear picture for Romania's future was emerging. Romania and each of its fellow Communist countries, satellites to the USSR, seemed to have its own logic, its own brand of politics that was impenetrable to an ingénue like myself.

The greatest changes I experienced, though probably quite unrelated to Operation Danube, was that two colleagues in the University, quite independently, showed a willingness to befriend me.

Dino Sandulescu

"Professor Mackay, please excuse the awkward timing. I am a fellow-professor in the School of Foreign Languages. My name is Constantin George Sandulescu, but you may call me Dino." He addressed me in a modulated, self-assured voice that any BBC announcer would covet.

Dino was slim, some years older than me, distinguished in a formal, double-breasted pin-stripe suit. Even with his academic stoop he towered over me. His spectacles drew attention to his eyes. They were wise, lively, amused and unafraid. In the impeccable English used by all my colleagues, he expressed his pleasure at meeting me and suggested we have a coffee together. Since I spent most of my time alone, I was more than willing to hear someone say they wanted to spend time with me. Moreover, this year I'd decided, I would respond to every overture that came my way.

After classes we met in a busy coffee-shop on the main boulevard. It was a coffee shop I avoided because, through the open door, I often saw colleagues inside and had discovered that if I entered, they invariably excused themselves and left. Having no wish to compromise any of them, it was simpler to avoid inflicting my company on them.

Dino and I sat at a table sipping Turkish coffee and conversing as if we were already old friends. He explained that we had not met before because he had spent the previous two years in England. It occurred to me as he told me about his literary interests and the dissertation he'd just written on Joyce's Finnegan's Wake at Leeds University, how much I missed simple, informal, straight-forward human contact and conversation.

That hour's conversation showed Dino to be a highly educated, well-read intellectual of a kind I had seldom met before arriving in Romania. Not given to small talk, he brought up and pursued profound and complex matters involving literary criticism, philosophy, communication theory and cybernetics. His mind moved rapidly. He frequently stopped in mid-sentence leaving me struggling to complete his thought. I was flattered that he appeared to assume that I could follow his multiple lines of simultaneous thought, his sudden digressions and uncompleted fragmentary sentences.

Despite a good degree that had covered English and Spanish languages and literatures, moral philosophy and general psychology, I was far less well-read than Dino and many of my Romanian colleagues. Like many educated Romanians, he preferred to read literature in the language it had been written in whether that be Romanian, English, French, Italian, or German. Dino was an educated person rather than a person who had acquired an education. In Scotland, at least the Scotland I knew, we talked of my generation 'getting an education' as if I were somehow myself first and education were a mantle that I could add on top of me. Many Romanians I had encountered since my arrival in the country, struck me as essentially *educated people* rather than simply persons who had donned a superficial cape of education over their being.

I immediately liked Dino. I enjoyed his enthusiasm, his willingness to sit with me and converse about matters important to him in a busy cafe and pay me the compliment of assuming they were also important to me. Before we parted, he wanted to know exactly where my apartment was and so I gave him the address and described the intersection where my building stood. He listened carefully.

"What are the most distinctive shops close by?"

"There is a Gospodina on the ground floor of my apartment block," I told him, a shop that sold ready-made food. "There's also a baker's shop across the road" On the occasions that I could find it, I bought "pâine Graham" there, bread as close to a whole-wheat loaf as I was able to find in Bucharest. I liked "pâine Graham" for the additional reason that it bore a familiar Scottish name!

Dino took my telephone number. "I will call you. I want to show you a translation of 'Giacomo Joyce' that I've just completed." He assumed I knew who or what Giacomo Joyce referred to. The truth was I did not.

Unusually early next morning, my telephone rang. "Bună dimineața Profesor Mackay!" I immediately recognized Dino's clear voice. I started to greet him but he intentionally spoke over me. "Ai nevoie de a cumpăra pâine Graham. Acum!" "You need to buy a loaf of Graham bread. Right now!"

I was used to reading between the lines when Romanians spoke. There was a concealed message for me, but what was it? When a Romanian sent you a message you either got it or you did not. If you did, you were elevated in their estimation. If not, your capacity was forever suspect.

"Acum?" I repeated, playing for time. "Now?"

"Imediat!" He insisted. "Immediately!" He hung up.

The penny dropped!

Quickly I donned shoes and coat, left my apartment, walked down the stairs and stepped out onto the street. At the corner, the lights took forever to change from red to green. With virtually no private cars, there was seldom any traffic, but Romanians did not jay-walk and I had learned that if I behaved in every way like a Romanian, I drew less attention to myself.

On green, I crossed Bălçescu and walked the half block to the baker's shop. Dino was wearing raincoat and his ubiquitous navy-blue beret. He was looking at the empty window but did not raise his head. When I was within twenty paces of him, he turned and began walking to the trolleybus stop and waited. I caught up to him, but he gave a barely detectable shake of his head. I was not to approach him. He let the first trolleybus pass and boarded the second. Allowing passengers between him and me, I followed. In a loud voice he asked the passenger standing next to him, "Comrade, does this bus go as far as Cișmigiu Gardens?" and got the reply, "It does comrade!" So that's where we were going -- to the beautiful Cișmigiu Gardens that I knew so well and enjoyed so often.

At a discreet distance I followed Dino's distinctive raincoat and beret round the gardens, past the seated chess-players. Eventually he sat on an empty bench, raised his head, indicating that I could join him.

"Buna dimineața, Profesor Mackay!" "Good morning" He drew a thin, soft-covered folio out of his briefcase, explaining in Romanian as he did so, that this was a galley proof version of Giacomo Joyce about to be published by Faber and Faber. Joyce had written though never published the sixteen-page manuscript, more than fifty years earlier. It is a love poem in which he tries to enter the mind of a 'dark lady' with whom he had apparently had an extra-marital affair. Dino was enthusiastic over this about-to-be-published posthumous work. He had obtained the printer's proofs before he left England a few days previously and had immediately sat down to translate the entire work. He had already extracted a promise from the Romanian State press that they would give priority to publishing his translation.

"My text and the Faber and Faber edition of Giacomo Joyce will appear simultaneously! My Romanian translation will beat all other foreign languages to the press!" He was proud of his coup!

He made no mention of his call, the coded message or the precautions he had taken. I knew that if I'd failed to understand his message or to follow his precautions, Dino would have written me off. That was how things worked in Romania under Communism. Had I not succeeded in learning these most important lesson during my first year in Bucharest, Dino and I would never have developed the profound trust, respect and solid friendship that began in September 1968 and continued to his final days.

I could not have imagined the adventures we would later share in Sweden, the UK, Canada, and Monaco.

Harald Mesch

Harald Mesch, also a colleague at the University, approached me cautiously. He waited until I left the building, followed me to the corner of Rosetti then "accidentally" met me on the corner of Magheru.

About 27, Harald was Germanic in looks with fair hair and an athletic frame. He stood several inches taller than my meagre five-foot-seven!

"Professor Mackay. Can we talk?"

Since he, too, had shown the initiative, I was delighted to make time for him. I suggested a restaurant where I occasionally had lunch about half a kilometre in the general direction of my house. Never had I bumped into any faculty member there nor seen anybody I knew. I'd tasted my first plate of çiorba in that restaurant and went back sufficiently often for the waiters to recognise me and deposit a menu on my table without my having to beg several times for one to be brought.

Service in Romania was notoriously poor. I put it down to poor training and that the concept of customer service barely existed since all employees were public servants. Poor service was also the case among public servants in post-War Scotland, the Post Office being one of the worst. Public servants seemed to be under the impression that they fulfilled their duties by turning up to work and that nothing beyond their surly presence was required, certainly not an obliging manner or an offer of service.

Harald and I talked about our university teaching. He chose his words very carefully and listened with equal care. I was used to being assessed, checked out and scrutinised. It was what Romanians did when they first met me and I appreciated the reasons, and the need, for their circumspection.

His specialty was 20th century American poetry, E. E. Cummings, Wallace Stevens, Carlos Williams, Carl Sandburg and Robert Frost. My courses at Aberdeen University hadn't included American poetry but I had read some for myself and had emigrated to the USA in 1964 when Robert Frost was made Poet Laureate. I told Harald I loved Frost's work. That set us off to a good start.

He told me that he was married and that his home was in Hermannstadt, a Saxon town in Transylvania. When he described its location, I realised I knew it by its Romanian name, Sibiu. I knew that name for two reasons. The hikers I met in the Buçegi talked enthusiastically about the Munții Făgărașului, the highest mountains in the Southern Carpathians. The place to start from, they agreed, was one of the villages on the road between Brașov and Sibiu. They talked of peaks 2,500 metres high on the southern flank of the Olt Valley. The mountains bore romantic names - Moldoveanu, Negoiu, Urleanu and the resounding Vânătoarea lui Buteanu.

The other reason I knew Harald's hometown was because it was the principal town north of Râmnicu Vâlcea in the Olt River valley. On two occasions the previous year I had visited and stayed in the one of the few functioning Romanian Orthodox convents, Horezu, just east of Râmnicu Vâlcea and I planned on taking my mother there when she came for her second visit in 1969.

Harald was a proud Saxon. The colonization of Transylvania by Germans, he told me, began in the 12th century. King Géza II of Hungary invited Saxons to settle in Transylvania to defend the south-eastern border of his Kingdom. The colonization from Germanic lands to the west, continued for over a hundred years. The colonists spoke Franconian dialects and these dialects still survived. In addition to his 12th century Franconian German, Harald also spoke fluent High German, Romanian, Russian, English and French. Most educated Romanians regarded multilingualism as a natural part of education and scorned those who spoke only a single language.

"I would like to show you Hermannstadt, Professor Mackay." Harald promised to let me know when a suitable weekend for him might be. I received invitations so infrequently from Romanians, for obvious reasons related to their self-protection, that I accepted his promise with undisguised enthusiasm.

Domnul Zamfirescu

On one occasion in 1967, instead of waiting for an invitation to be issued, I had made the reprehensible mistake of thoughtlessly inviting myself on a day trip with a Romanian, Domnul Zamfirescu. That embarrassing experience taught me an important lesson.

The gentlemanly Domnul Zamfirescu was the official translator at the British Embassy. With his dignified bearing and perfect upper-class English accent, I mistook him for the British Ambassador when we first met. Well-dressed, with a pleasant expression on his tanned face, he invariably greeted me warmly and made a few minutes conversation if I bumped into him on my occasional visits to the Embassy.

He invariably asked about my excursions into the mountains. He too, had been a keen outdoorsman in his youth and now his passion was for fishing. He made regular trips with rods and lures to the shallow lakes that drained the Argeş River just a short train-ride to the north-west of Bucharest.

I'd been in the country perhaps two or three months and was feeling frustrated by my failure to make close contact with Romanians other than Karen, my informant for the Secret Police. So, one day, when Domnul Zamfirescu greeted me in the Embassy and told me he was going fishing that weekend, I seized the opportunity.

"May I come with you?" I asked.

The smile left his face; he was silent for a moment. "Yes." He soberly agreed. "Meet me at the train station before six on Saturday morning. Dress warmly. I will bring rods, you bring sandwiches." I could tell that he wasn't overly enthusiastic.

Before six, I was waiting for him at the Gara de Nord. He was dressed in heavy wool trousers and an old, but beautifully tailored, tweed jacket. Separately, we bought tickets, boarded the train, and sat together in a vacant compartment. An hour later we alighted at a tiny village on a desolate plain under a leaden sky. We began to walk.

Within half an hour we arrived at a cluster of wooden sheds round a pier on the shore of a lake. I could see a score of wooden cobbles tied up to the pier. Domnul Zamfirescu exchanged polite greetings with the man in charge of the boats and we were helped into one.

"Can you row?" He asked me.

"I can." I took the oars while he sat in the stern arranging his fishing rods and tackle. He gestured in the direction he wanted to go.

As children we had learned to row small boats on Scottish lochs. As a student, I'd worked as a grouse-beater on Highland estates and often spent the weekends with one or other of the gamekeepers shooting rabbits and occasionally fishing from a rowing boat. With strong uniform strokes I headed into the cold wind. By the looks of things, it would blow all day. When we'd reached a point he liked, he asked me to ship the oars. We drifted on the broad expanse of water. He handed me a rod with a spinning reel and a lure on the line. I began to cast from the prow while he cast from the rear.

"What are we fishing for?" I asked.

"Pike!" I was delighted. The ghillies I'd worked with in Scotland told me how fierce they were.

We rowed and cast, rowed and cast for a couple of hours without a single bite.

He suggested a break. I produced sandwiches and a flask. With our backs to the wind, we warmed ourselves.

"I admire your jacket," I told him. "It could almost be Harris Tweed." His face lit up. He undid the buttons and showed me the label inside. It bore the characteristic orb, the symbol given by the inspectors of the Harris Tweed Authority to tweed hand-woven by islanders in the Outer Hebrides of Scotland.

"It is real Harris Tweed!" I showed my astonishment. He didn't disguise his pleasure.

As we ate, he talked. In a very matter-of-fact way, he told me that his family had owned large estates in the region of Maramureş in northern Romania. His family had maintained their estates, their many employees and their home in the capital, by logging their forests, milling the lumber in their own sawmills and selling the boards to the construction industry. From the age of 12 until he was 18, he had been sent by his parents to attend a minor public school in England as a boarder. Later, he had studied economics at London University and, when he returned to Romania, became the manager of his family's lumber business in the forested valleys of his beloved Maramureş. After the Second World War, the Communist Party took over the government, the family estates were confiscated and those who had had businesses and employees were branded "exploiters of the people". He had had difficulty finding work until he had been assigned, he was vague on precisely how this had come about, to be the governmentally approved interpreter and translator to the British Embassy.

"My Harris Tweed jacket was made for me in Saville Row in 1935!"

I was pleased to have recognized it and to have given him the opportunity of telling me its history.

From time to time we saw other boats. Then I became aware of one of these with a single man at the oars. It appeared to be closing in on us over the choppy water. Domnul Zamfirescu's attention was on casting, so I felt I had to draw his attention to the approaching coble. When he saw, he appeared disconcerted.

At a hundred yards, the rower turned and hailed him by name. Domnul Zamfirescu returned the greeting awkwardly, I noticed, and without enthusiasm.

"Yesterday you cancelled our regular fishing trip. Now what do I find? You, fishing with someone else!" The speaker called in Romanian. He was a gentleman similar in age to Domnul Zamfirescu. He eyed me curiously. Domnul Zamfirescu made no attempt to introduce me and the gentleman made no attempt to introduce himself.

"My sincere apologies!" Domnul Zamfirescu gestured in my general direction as if my presence were all the explanation needed. The boatman gave a penetrating look, nodded, and rowed off to fish alone.

"My friend and fishing companion of many years. I called him yesterday. I told him I was unwell and couldn't make it. I thought he would cancel."

I felt Domnul Zamfirescu's desolation and spent the rest of that raw, invigorating day feeling guilty. I swore never again to impose my company uninvited on any Romanian. That was a sound resolution that served me well in that complex and perplexing country where nothing was ever as it seemed.

Parkul Herăstrău, one of the many beautiful parks that graced Bucharest.

27

AND MORE

Dudu Popescu

Dudu was an engineer in his mid-30s, with a ready smile but an air of fatigue and resignation. I may have met him and his girl-friend Gabriela, Gabi for short, in Sinaia on one of my walking excursions in the Prahova Valley. We became friends.

Dudu was one of the few Romanians who would permit me to speak exclusively in Romanian. Most others would switch to English the moment I got into difficulties. Dudu put up with my defective Romanian because, as a speaker of only Romanian and Russian, he had no option. We enjoyed each other's company. He appreciated my discretion and that I loved the city of Bucharest with its elegant buildings, churches and parks. He, Gabi, and I often walked in these parks, watched the chess players, or sat each with a glass of beer and a plate of gherkins and cheese at one of the outside tables by a lake.

Dudu was the most open and communicative Romanian I met during my two years in the country. It seemed as if he understood how detached my life in Bucharest was from normal human society and had decided to make up for that by showing me how the average Romanian lived. He invited me to his apartment in a block of flats not far from mine. His block stood directly on the main boulevard that I took daily to the University. His was a tiny bachelor apartment, what I would have called a bed-sitting room, no more than four metres by six metres. There was a mattress and a bookcase, two chairs, and the space to the left of the door was divided into two by a wall that separated a minute sink and cooking area from a tiny bathroom with the shower-head directly over the WC.

Dudu showed his sense of humour by laughing uproariously when I commented that his bathroom arrangements proved that Romanians were more efficient than the British. "You can enjoy all of your morning's activities in the same place and simultaneously!"

He told me that he'd been an engineering student at the time the USSR successfully put the first satellite into orbit in 1957. The achievement was widely talked about around the world and of course in Romania. A fellow student initiated a discussion about the relative scientific capabilities of the USSR and the USA asserting that Sputnik proved the USSR was superior. Dudu expressed the view that the USA was just as, if not more capable, and might soon better the Soviet Union's triumph.

It turned out that the student was an *agent-provocateur* for the Secret Police. He informed on Dudu. Dudu was tried, found guilty and interned in a labour camp for two years. Several years after his release, he was allowed back into the engineering programme. Since graduating, however, his career had been hampered.

Like most Romanians I met, Dudu assured me that I was continuously under State surveillance and my apartment and telephone were bugged. He developed a set of procedures should we want to contact each other. I had to use not my own but a public telephone in the street. I had to learn to slip several fingers into the rotary dial at the same time. Watching agents, he warned me, were trained to identify the number I was calling from across the street if I dialled using a single digit. If he was home and picked up, I simply gave him a number. That number had to change frequently and represented, when multiplied by two, the number of minutes within which I would knock at his apartment door. If he could accept my visit, he doubled the number I gave him; if not, he uttered the single word, "Nu!" We used the same procedure to contact me at my apartment. We could never enter either his building or mine together. One of us must enter alone; the other circle the block and make his entry independently.

Dudu and Gabi liked to visit my apartment probably because of the larger space they were free to move about in. I cooked meals that they enjoyed. The fish or meat or chicken, I bought, as they did, at the peasants' market but I had spices like curry that they'd never tasted before, comfortable chairs and a table to sit at. They liked music. I'd brought back with me from the UK a radio-cum-cassette-tape-player and some Beatles' tapes. *Hey Jude* had recently been released. We would have a glass of Johnny Walker from the British Embassy commissary and spend the evening laughing and talking in the darkening apartment. Dudu had a private source who supplied him with fragrant white wine and, occasionally, I would trade him two bottles of whisky for a demijohn of that wine when he was able to get it.

My mastery of Romanian and my social life improved immeasurably because of my friendship with Dudu and Gabi.

In the early spring of 1969, Dudu announced that he had procured a ticket for a State-organized tour to Istanbul. The Romanian Government was under pressure to show the West that it was not a prison and had begun to organize tours to other Eastern European countries and even to cities in countries outside the Warsaw Pact group such as Paris and Istanbul. It was extremely difficult for a Romanian to purchase a place on such a tour and clients were on a waiting list for months before being accepted or rejected without any explanation. The cheapest trip was to Istanbul, just a relatively short flight to the south-east.

Such trips to the West were not without their risks for the Romanian Government. Every week the CIA-funded propaganda station "Radio Free Europe" would read the list of the names of Romanian nationals who had defected and been granted political asylum. Both the Romanian and the Soviet Governments jammed the broadcast but so ineffectively that I, with patience, could pick it up on my radio.

As the date of his tour approached, Dudu's excitement increased. Now he had the ticket, he told me, and with ticket in hand he could apply for a passport. He applied for and obtained his passport. For a couple of weeks, we didn't see each other. If I called his number from a street phone and gave the code "fifteen" or "twenty" Dudu invariably replied "Nu!" I'd adopted the capability of accepting odd behaviour without having to understand what lay behind them. I hoped I might see him before he left on his ten-day trip.

My phone rang. Dudu's voice. "Five!"

"Ten!" I answered. And Dudu was at my door minutes later. I opened the door, but he gestured me to follow him and left. I put on my shoes, closed the door and walked down to the street. Dudu boarded a concertina bus. I followed. He alighted at a park. So, did I. After a circular walk he sat down on a bench. I joined him. He came straight to the point.

"You know I leave for Istanbul with the tour tomorrow evening."

"I do."

"I've spent the last while selling everything I own." Dudu looked at me his face impassive.

He wasn't coming back! I merely nodded.

"I'll request political asylum at the American Embassy in Istanbul. My request will be accepted. I'll be sent to a camp for displaced persons inside Turkey. With a bit of luck, I'll be admitted to the US after a year or two." A meticulous Romanian, he'd exercised due diligence.

Then Dudu looked at me with pain in his eyes. "I dare not tell Gabi." I knew exactly what he meant. "I want you to explain to her when she finds out!"

"Me?"

"You can make her understand, Ron."

I felt some of the hurt he was feeling. In a Communist country run as a prison, where you never knew who might be an informer, you dare not risk confiding in even your lover!

"I will, Dudu!" We shook hands. He walked away. I watched the forlorn figure fail to board a crowded trolley bus and resign himself to a lonely walk back to his tiny, now empty, apartment.

For the next few days, I listened for Dudu's name on Radio Free Europe. The CIA was thorough and milked every ounce of propaganda they could from broadcasting the list of defectors from Central and Eastern European countries to the West – or in this case to a US embassy in Turkey. I was concerned that something had gone wrong.

Five breathless days, I waited. Then from a street phone, I rang Dudu's number using multiple digits to dial. I let the phone ring. In my mind's eye, I could see it sitting on the empty floor. It rang and rang and rang. "Is that good news or bad?" I asked myself. Suddenly it stopped ringing. It was picked up but not answered. So surprised was I that I just looked at the silent mouthpiece; I'd had no plan before I dialled.

"Ten!" A single word would be enough. If it was a Securitate agent who picked up, he wouldn't respond appropriately.

"Twenty!" It was Dudu! He was still in Bucharest!

I suffered ten long, painful minutes striding to his apartment building before knocking. Dudu cracked the door open just far enough for me to see a grim, sleepless face. He gestured that I should leave. I went down to the street. Minutes later, Dudu appeared. At a distance of 50 yards I followed him to a park and after a circular walk he sat at one end of an empty bench, head bowed. I sat down at the other end. He raised his eyes and looked at me, disconsolate. I said nothing.

"I got as far as Otopeni Airport." His voice was hopeless. "I joined the tour group. We passed through immigration and customs. Then, out of the blue, came a final security check. They searched each of us thoroughly. I was carrying my Romanian address book. They confiscated it. Securitate was called. Uniformed agents told me I was carrying forbidden material. I showed that all it contained were the names and telephone numbers of my family and friends in Bucharest, so I could send postcards. Nevertheless, they detained me, refused to let me board. After the plane left, they sent me home."

"Why?" I was flabbergasted.

Dudu shook his head slowly as if taking the blame on himself. "The night before I left, there was a knock at my door. Very late. After 11 o'clock. I didn't answer."

He paused, still shaking his head in disbelief. I didn't dare interrupt.

"The knocking persisted. What could I do?" He paused again. "When I opened the door there was a civilian standing there with Securitate ID in his hand. He wanted to talk to me. I agreed, prepared to talk right there on the threshold. 'This is a confidential matter,' he said. How could I refuse?"

His eyes were pleading; I nodded my silent understanding.

"The Secret Police agent made no comment about my apartment being empty. I'd sold whatever I could sell and had given away the rest! Only my suitcase stood ready in the corner. The telephone was on the floor. For obvious reasons I hadn't cancelled it. We stood there, the Agent and I. He was first to speak. 'When you reach Istanbul, Domnul Popescu, we would like you to watch your fellow tourists throughout the trip and report any anomalous behaviour to us on your return. Agreed?'"

"What did you say?" I immediately bit my tongue for asking such a foolish question.

I could see the defeat and the shame deep in his eyes.

"He told me how to contact him when I got back. Then he left. I couldn't sleep, and it wasn't just because I was lying on the floor! The following afternoon I presented myself at the airport." He shrugged. "You know the rest."

There was nothing I could say to alleviate his misery.

"Now I'll never again, never, be given permission to leave!" He looked desolate. "I still have my job. I still have Gabi!" He tried to smile. "I just have to buy my furniture back before Gabi finds out what I planned to do. Nobody knows I'm still here. Except you." He trusted me.

I offered Dudu all the Romanian Lei I had. He accepted. I returned to my apartment to fetch them.

28

ROMANCE

One morning after my last class of the day, I left the University to walk home alone as usual. I noticed that 'M', one of my best and brightest students, fell in twenty paces behind me when I left the building. There seemed to be an unspoken rule that forbad my students walking with me or I with them in public, so I thought nothing about it. We were simply heading in the same direction. However, I was a brisk walker and 'M' had to be adjusting her pace to sustain that fixed distance behind me. To doublecheck if our directions coincided by mere coincidence, I turned off the boulevard onto Strada Batiştei. Four or five minutes later, 'M' was still behind me.

The side-street I'd chosen, though central, was quiet, had been elegant but like much of the city had fallen into disrepair since the Communists took over. I stopped. 'M' caught up with me and smiled.

"I wanted to speak with you, but not in the University nor on the main boulevard." No further explanation was needed. Her English was excellent. She had an open, intelligent face and clear, warm, alert eyes that twinkled with amusement as if the pursuit of privacy were an obligatory but slightly infantile game. Cheeks flushed from the exertion of keeping up with my habitually fast pace, 'M' positively glowed with health. She still carried just a little late-teenage puppy-fat.

In class, 'M' was an absolute delight. In addition to being attentive, highly intelligent, and usually a step ahead of most of her fellow-students and sometimes of me, she had a quick wit and her humour tended towards the lightly satirical. She possessed the skill of delicately mocking the cozy half-truths uttered by the more orthodox, less insightful students. Despite her being one of the most naturally able students, her companions liked her. She never tried to score points at the expense of others; she was a true team-player and a natural leader. Although in her late teens, she possessed a wisdom and depth far beyond her years. Like her fellows, she was cleverly playful and her acute, under-stated observations would have all of us laughing out loud. She was popular because she was effortlessly outstanding, never thrusting to impress.

Together, we continued to walk away from Magheru Boulevard. Although the side-street was quiet, walkers drew less attention than couples standing still.

"I love your classes. We all love your classes, Professor Mackay!" I thanked her. My classes were the social highpoint of my week. Besides providing me with pleasant human company, my students educated me about the lives of young Romanians and how they spent their time. They were not unlike we had been in our first year at Aberdeen University. Our studies came first, our hobbies second and our social lives limited to a few hours at the weekend. A natural discipline that served everybody well.

'M' asked me where I had studied and how I had learned so much. I was being flattered but thoroughly enjoyed it, and offered her a little of my background: two years of experience working and travelling in Europe, Morocco and the Canary Islands before I went to university; my third year of studies at the University of Madrid; arriving at Idlewild Airport in New York as an immigrant; renouncing my green card a few months later because I'd been obliged to register for the draft and would have been sent to serve in Vietnam had I stayed; then back to Aberdeen to complete my degree.

To my surprise, 'M' knew quite a bit about the Vietnam War because she had a cousin in California. He was slightly older than herself and would have to serve in Vietnam unless he won an exemption by attending graduate school. Her knowledge encouraged me to go into a more detail of my encounter with the draft board. In 1964, the Americans were giving their last though ultimately futile thrust to win in Vietnam. I presented myself, as demanded by the official letter, to register at the military office in Massachusetts. Hoping to win a reprieve, I explained to the overweight major in charge that I had already done military service in a special unit of the British Army.

"Special training?" Her eyes glinted.

"Winter warfare." She looked blank so I added, "In small, self-supporting groups, our role was to create havoc behind enemy lines without being caught." Her eyes lit up.

"Weapons?"

"I'm proficient in weapons used by the British infantry. Rifle, automatic weapons, rocket launcher, mortar, grenades, hand-to-hand combat–" She interrupted me.

"That's exactly why we want you in Vietnam, Sir!" She barked a laugh and stamped my papers.

"And that's exactly why I'm going to leave the US, Major!" I thought. I left a few weeks later.

As we walked and talked, 'M' explained the reason for having pursued me.

"'D' wants to invite you to have tea with us one afternoon next week."

"'D'?" I was puzzled.

"My mother. I call her by her first name. She pronounced a Christian name I recognised as Hungarian.

My surprise that 'M's mother wanted to invite me to tea must have been apparent. 'M' continued.

"My mother trained as a concert pianist. She used to teach at the conservatory here in Bucharest. Now she works for Romanian State

Radio. At home, she offers master classes to pianists prior to their giving major recitals here or abroad." This information didn't help me understand why 'M's mother might want to invite me to tea, but I knew better than to ask directly.

"None of the students and none of the professors should know," she advised.

"Your mother's invitation sounds delightful," I'd promised myself not to miss any opportunity offered me to socialise with Romanians. "I understand the need for discretion. I'm free every afternoon."

'M' smiled her pleasure and returned towards Magheru. Mulling over what had just happened, I walked slowly back to my apartment.

In Romania I'd quickly learned that there was nothing for nothing. All encounters were commercial transactions of one sort or another. There was always a *quid pro quo*. I had no objection to friendships as exchanges, "I offer so that you will offer", so long as the objects of trade and the rules of the game were transparent. To their credit, Romanians favoured a win-win approach; they understood the importance of both parties ending up satisfied.

I couldn't fathom why an illustrious Hungarian-Romanian musician would want to invite me to tea.

The following week, 'M' again tailed me and caught up with me. She caught up with me on the side street.

"Our appointment for tea is set. Memorise our address but don't write it down." 'M' explained. I would find her waiting for me by the gate bearing the house number at precisely four o'clock on Thursday. I was excited and invigorated at the thought that I was being invited into the real lives of Romanians.

At a quarter to four the following Thursday, I was dressed, shoes polished, and buying flowers in the same florist's shop in Piaţa Romana as the day I had visited Madame Cartianu some 14 months earlier. To the florist's delight I asked for 11 roses. She congratulated me on my improved command of Romanian. Then I had her wrap a 12th rose separately.

"For your sweetheart's little girl!" She nodded knowingly.

"You guessed correctly!" I congratulated her insight. I was flattered that she thought I might have a lover. In Romania, judging from the stories my students told me, to be without a lover was to lack an essential element of life. "Aventuras" – casual encounters – were regarded as the next best thing. These intimate confidences surprised me but I had been truly shocked when my students told me that it was not uncommon for a female student's final examination result to be withheld until she had visited a certain professor at the villa belonging to the Writers' Union in Sinaia. The horror on my face simply encouraged gales of laughter and so I wasn't sure of the truth of that assertion.

Flowers in hand, I crossed the Piața wondering what lay in store for me at 'D's tea party.

Their house was in a quiet street, set back from the road. It looked grand but, like most, in need of maintenance. There were wrought iron gates: one large that appeared to have served as the carriage entrance and a narrow gate for people.

As I approached, 'M' opened the smaller gate, gave me a huge smile when she saw the roses, and put her finger to her lips. "Silence!" I was used to being asked not to speak. My accent attracted attention and no Romanian wanted to be identified as associating with a foreigner.

We approached the service door. I could see that this had indeed been a grand home thirty years earlier. A mature cedar tree gave sobriety to the building surrounded by small untended lawns and gardens. 'M' ushered me through the service door into the ground floor corridor and again signalling silence, led me up a set of stairs, along another corridor with high ceilings and paused in front of a tall, beautifully-panelled wooden door. "Here!" she indicated, silently.

Closing the door firmly behind her, 'M' ushered me into a very spacious and elegant room with a set of tall windows on one wall and an alcove in another. The remaining walls were taken up with bookcases, paintings and framed black-and-white photographs. The warm hardwood floor was well-worn. At one end stood a polished grand piano and at the other, an equally well-polished upright. The high ceilings bore decorative plaster mouldings. Under the chandelier,

'M's mother stood, poised, elegant, beside a table bearing china tea set and polished silverware.

She greeted me in both Romanian and English and asked me to call her 'D'. She was a slim, good-looking, woman about the same age as my own mother. Her warm intelligent eyes missed nothing. She offered me her hand in a way that suggested I should kiss rather than shake it. I had the presence of mind to fulfil her expectations. It was not uncommon to see a well-dressed man kiss the hand of a well-dressed woman at a concert or at the theatre. That simple gesture of respect was the cultured greeting I'd see occasionally on Magheru Boulevard. Handshakes were more common at work and in political gatherings.

When we meet a person for the first time, we instantaneously rely on innumerable cues to help us to decide who we think they are. We use context, gender, age, dress, bearing, speech. We draw on what we already know about the person, their training, profession, position, hobbies. We use their voice, pronunciation, bearing, their demeanour and the light in their eyes.

Immediately I appreciated that 'D' was someone used to a refined, intellectual world of values, culture, and beauty. The kind of person whose quality of life, while fortified by material possessions, is neither conditioned by nor entirely dependent on them. Hence, if some or most, or perhaps even all, of the material props are taken away, confiscated or removed, these individuals find ways to retain the rich texture of their lives and continue to live with dignity and self-respect. They can draw on inner resources, as well as on the beauty of nature, music, paintings, literature, friends and conversation. The depth of their minds ensures that their spirit is nourished and in turn sustains everything around them, home, family, workplace, and those they come into contact with.

'D' was undoubtedly that kind of fortunate person.

Switching from Romanian to French, then English and back to Romanian, she talked briefly about her work as musical director with the State Radio. With greater enthusiasm she told me about preparing musicians for major public performances. A little, a very little, about her background.

"I am from Timişoara, a Hungarian-Jewish family. In Budapest I studied piano at the conservatory. My husband was also a musician. Later I was appointed to teach at the conservatory here in Bucharest. This," she gestured widely to embrace the whole floor, "was our home. Then came the War. The War ended and the Communist Party took power. My husband," she paused, "My husband died. I lost my position in the conservatory. Now I am musical director for State Radio." She shrugged as if the waxing and waning of personal misfortune were normal.

'D's voice was dispassionate; made no bid for sympathy. The voice of a proud survivor.

'M' poured tea. 'D' offered me a pastry layered with black poppy seeds, plum jam, crunchy walnuts and sliced apples and then *profiterole*, tiny chocolate-covered pastries filled with cream. I drank tea, ate as little as was polite, lacking a sweet tooth, and listened, entranced.

'D' and 'M' were more than mother and daughter; they were best friends, peers.

"How do you find my daughter, Professor Mackay?"

I found the question embarrassing with 'M' sitting right there with us at the table. To my relief, 'D' didn't wait for me to respond. Instead, she listed 'M's talents. Some I had seen for myself, many I had not. My nodding satisfied her. 'M' smiled openly, calmly acknowledging the truth of 'D's faith in her.

"And you?" I understood that what 'D' wanted to hear about in particular was my academic and intellectual pursuits. They were modest compared to her daughter's but I'd had a thorough grounding in English literature from Beowulf to the post-war poets, knew British drama and the contemporary theatre well, had read widely in English and in Spanish and had experience of a variety of cultures. These credentials along with my position as Exchange Professor, although I ran into a little difficulty trying to explain that the post my counterpart from the Bucharest University had assumed in Cambridge was not in fact mine, appeared to satisfy 'D'.

"Success for a student in Romania depends on being the very best!"

'D' made an assertion I could readily agree with, knowing by now how competent and competitive students were. "I would like you to tutor 'M' in English literature using only the English language so that she can improve in both simultaneously." 'M' smiled encouragingly at me. So, this was the purpose of the invitation!

"Of course," 'D' added, "I will pay your fee."

By the time I left their home as quietly and as unobtrusively as I had entered, for this elegant room with the alcove which contained their pianos as well as the adjacent bedroom and kitchen constituted their entire home, I had to my surprise, agreed to serve as M's personal tutor. Two hours every second week. Because the weather was still good, we agreed to meet at an outdoor table by the restaurant on the lake in distant Herăstrău Park at the terminus of the tramline.

As I walked back to my apartment, I puzzled over several matters. How had I allowed myself to be persuaded so easily? How cavalier had I been in dismissing the matter of a fee? Above all, why didn't I regret what I'd agreed to do?

I failed to reach any satisfactory answers. A few days later, on a bright sunny afternoon, I rode in one of the two linked cars of the tram on its way to Herăstrău Park. 'M' was riding separately in the other.

'M' at seven months old. Smart, courageous and beautiful even then.

29

WHY WALK?

I walked to and from my apartment to my University department on Pitar Moş in the heart of the city. Public transport was frequent and inexpensive but there were three disadvantages. The first was that you had to engage in three separate fights for every journey you took. The first was a hand-to-hand, sometimes a knee-to-groin contest to board the vehicle. Romanians, from the youngest child to the most aged crone, were experts at using the sharp parts of their anatomy. Once successfully boarded, immediately came the second trial of finding a space large enough to breathe in. Passengers were officially encouraged to board at the rear then gradually move up the aisle and leave from the front. In practice, however, passengers crowded to either one end or the other so that when they wanted to leave, they were close enough to the doors to leap off before the human tide swept you back on again. Bus-stop behaviour in Bucharest wasn't driven by my British logic that you wait until passengers alight before attempting to board. Pitched battles were the norm.

To encourage riders to move towards the centre of the bus, the conductor would keep up a steady chant.

"More to the middle. Move up please!"

Nobody paid the slightest attention, except me, that is. I didn't care to have my face squashed into the back of some smelly old coat that had seen better days. Dry cleaners were unknown in Bucharest then, and if it had been raining, the smell of wet wool combined with stale sweat was overpowering. The odour grew steadily worse as winter progressed. My strategy was to move towards the less-crowded middle of the aisle where I could breathe. Many's the time I'd been carried past my intended stop.

Hence, I preferred to walk. The distance was a mile or a mile-and-a-half. Little traffic meant that the air in the city was fresh and besides, I never failed to see something to capture my interest on my walks.

Cold Turkey

Early one fine morning with an autumn chill in the air, I was striding to the University for my first class. A pale sun shone; the sky was blue. Near Piața Unirii, shoppers' bulging string bags reminded me it was market day. Bargain-seekers haggled with weather-beaten farmers for fresh vegetables, unblemished apples, the plumpest chicken, pickled cabbage, and the brightest chrysanthemums. Opportunistic peasants, open sacks in hand, accosted shoppers to offer a price below that of the stallholders. Even Communism can't extinguish the entrepreneurial spirit.

A squat peasant with a sheepskin waistcoat and a scarf round her head, held an obstreperous turkey by its feet. The turkey, as big almost as the woman was squawking indignantly at being carried upside-down. It struggled, slipped its cord, and leapt free. The old peasant, seeing a year's income skip off, let out a screech louder than any turkey and lunged determinedly at the bird.

The swell of pedestrians and more descending from full trolleybuses, stopped to roar at the spectacle of an ungainly peasant hobble after her rebellious turkey. Despite her bulk, desperation made her agile. She pounced, almost grabbing the bird, but it beat her with its wings and evaded her grasp. With a furious grimace and ferocious blasphemies, she chased the terrified turkey to the base of a tree. For several minutes, the crowd cheered her on as she vainly pursued the squawking turkey round and round the tree. Two men took pity on her, trapped the indignant turkey and slipped the cord more firmly round its legs.

The crowd cheered. Men raised their căçiulas. The peasant recovered her dignity and crossed Bălçescu on the green light, grinning in triumph. Noisily, the turkey continued to berate the world for injustice.

Snow-clearing in Winter

In winter, the streets of the capital were cleared of snow by brigades of Amazon-like women wielding shovels. One brigade would clear the snow off the pavement into the gutter, another brigade hurled snow over their heads into waiting trucks. The boards on the sides of the trucks towered above the women and they had to make a superhuman effort to sling the snow high enough for it to land in the truck-bed. If they miscalculated, the snow would hit the boards and cascade back down on top of them like an avalanche.

As soon as these brawny beldams loaded one truck, the driver took it to the bridge over the Dâmbovița River. Parking close to the edge, he lowered the sideboards and in short order, the packed snow was blasted off the truck-bed by a powerful jet of water from a pressure-hose mounted on a heavy tripod.

I'd watch these strenuous but effective efforts on my way in to work at seven-thirty. By the time I returned at midday the pavements and roads would be clear and drying in the warm winter sun.

A Good Samaritan

One late winter evening I was walking home from an evening class I taught voluntarily once a week at the People's University. Neither the students nor I had been inspired during the two late hours. They were tired. I had an incipient cold. Few had read the homework and I'd summarised it so clumsily that there was little enthusiasm and less discussion. Magheru Boulevard was dark, it was starting to rain and so, in my misery I crossed the street to catch a trolleybus home as my reward for an arduous day.

Everybody, it seemed, had the same desire to keep dry, so when bus after bus arrived full and left fuller, I decided to walk despite an inevitable soaking. Trolleybuses crammed full of damp humanity and people hanging out the doors kept passing me between stops. The rain fell steadily. As I arrived at the stop on the corner of Strada Ion Ghica, a trolleybus pulled up, its hydraulic doors hissing a welcome invitation. The only other passenger wanting to board was an old, very frail-looking lady dressed in black. I grabbed the door-rail with one hand and placed one foot on the lower step. With the other hand I grasped the old woman's elbow to assist her onto the step.

Just as I had her safely aboard, a tremendous blow to the back of my neck propelled me off the step. I landed on all-fours on the filthy wet street. Before I could regain my senses, the doors hissed closed. I watched the back of the old woman's head through the window as the trolleybus accelerated down Bălcescu spitting a shower of sulphurous sparks from the overhead wire.

I'd been in a poor humour to start with. Now I was really suffering and feeling sorry for myself. Here I was, working in a city where the law made it difficult for the locals even to speak to me, making my way home from a voluntary evening class where even my adult students were afraid to interact. My charitable nature had encouraged me to help a frail old woman onto the trolleybus, and what did I get in return? Ostracism and physical abuse! Even frail old women could deal a near-lethal blow to board a bus!

A strong hand grasped mine and helped me up. "Are you alright, sir?" I allowed myself to be steadied by his grip and looked at him, dazed. His concern gratified me and the fact that he'd called me "sir" appeased my hurt feelings. The usual address was "comrade" and I did not feel comradely! Still dazed, I looked at him, gesturing to the fast disappearing trolleybus.

"Did you see? That old lady hit me!" Though physical violence was commonplace at the bus-stops I was affronted. I'd assisted her with the best of Christian intentions.

"Ați fost electrocutat." "It wasn't the old woman. You were electrocuted, sir!" He helped me to the pavement. "Are you better now, sir?" Nodding, he left, worried about having spoken to a foreigner. "I'd been electrocuted?" I thought about it. When I grabbed the wet rail of the bus and still had one foot on the wet road, I must have acted as an earth to the electrical circuit. It wasn't an uppercut from the frail old lady's fist after all! With my faith in human nature slightly restored and to prove to myself that I was still alive, I began to walk the rest of the way. As I made my way down the empty boulevard, I saw the humour in my automatic assumption that the old lady had clobbered me.

Relieved that I'd escaped electrocution but still feeling sorry for myself and now feeling more poorly than before the accident, I waited at the bus stop by Strada Lipscani, the half-way point home. I paid no attention to the other people hanging around, too busy wondering how I could board the next trolleybus without touching it with my hands.

"Bună seara, Professor Mackay!" I turned at the greeting. There was one of my Bucharest University students, one half of a pair of identical twins. Both were in my class.

"I am Mihaela." Had she not told me, I wouldn't have known which of the two she was.

I returned her greeting and told her what had happened. She was sympathetic. I told her I had a cold coming.

"As soon as you get home, Professor, stir a spoonful of honey into a large glass of hot water and țuică, the local plum brandy, and go straight to bed!" Her concern and kindly advice made me feel better.

"In Scotland, we use whisky. We call the drink a hot-toddy!" I told her. "The trouble is, I have neither whisky nor țuică at home."

"Give me 5 Lei!" I did. "Now wait here." I waited. Mihaela entered a bar I'd passed many times but never entered. She emerged with a small bottle of țuică, just as the bus drew up. We both boarded.

We both got off the bus at the corner of Bălçescu and Mărășești. "This is where I live," I told her expecting her to hand me the țuică.

"I'll make it for you." She had decided to take control. "But nobody must know."

I nodded appreciatively. She could see that I understood the need for discretion.

When we reached my apartment, Mihaela told me to go straight to bed. Meanwhile, she went into the kitchen. Minutes later she brought me a large, hot glass. "Drink it all down!"

She watched me as I did, then took the empty glass back to the kitchen and I listened to her rinsing it. The last thing I remember hearing was Mihaela quietly closing the door of my apartment as she left.

When I awoke the following morning, it was at least an hour or two later than my usual time of 0530 but I felt renewed and on top of the world.

Walking to and from the University provided me with a range of experiences I'd have otherwise missed.

Because, in the country, Romanians were less afraid to have contact with me, I spent a lot of time outside the cities.

30
'M' AND 'D' COMBINED

Quality of Life

For those who have something thoughtful to say and the desire to say it, mastering a foreign language is easier than for poor conversationalists. 'M' was a thinker had interesting things to say, though never too much about herself, her family or other people. She talked intelligently about the books she was studying for her courses and what she was reading for pleasure. I was a willing listener and commentator.

I gathered that 'M' and 'D' had a vibrant group of intellectual friends who talked about the content of the books they lent each other, attended and discussed films and theatre together and always listened to good music. Perhaps because these good things of life were hard to come by in Romania, they were highly prized. The average educated Romanian certainly pursued, appreciated, and attended thoughtfully to all of these. In the West, I reflected, the range of cultural and popular options is so great that people can take them for granted, dismiss, and reject, disparage, ignore and even repudiate them. Not so in Romania.

The Claude Lelouch film *A Man and a Woman* was being shown in Bucharest. It tells the poignant story of widower and single father, played by Jean-Louis Trintignant, and a single, widowed mother played by Anouk Aimée. They meet through their children and form a friendship that turns into a romance. In Aberdeen or Edinburgh or London it would have been one of a choice of many films that people might see. They would certainly give it a miss if it were not in English. In Bucharest, where the acquisition of foreign languages was a social expectation, showings of this film in French were sold out. Those who had seen it discussed its every scene in great detail. 'M' discussed the film with me and recounted the discussions about it that she and her mother had with their friends without, of course, mentioning who the friends were. Every detail of the narrative, the drama, the characterisation and the photography were reflected on, assessed and critiqued. It was a nutritious bone to gnaw and to reflect upon with pleasure.

"The Fugitive", an American drama series, "Evadatul" in Romanian, was broadcast on Romanian TV every Thursday evening for weeks. Inexplicably, it was dubbed in Spanish as "El Fugitivo". 'M' said she and other students acquired basic Spanish from the series by listening attentively and reading the Romanian subtitles. On the evenings that The Fugitive was broadcast, there was nobody on the streets, nobody in the restaurants. Everybody stayed at home or visited somebody else's home to watch the American serial and then review and discuss it afterwards.

In the guarded, understated anecdotes 'M' shared with me, I earned a deeper insight into the vibrancy of intellectual life in Romania. She taught me how the essence of a penetrating anecdote could be captured and shared, while withholding personal details and thereby compromising no one.

We walked, talked, and held hands as the leaves changed colour. It was a beautiful time for us both.

'M' was forthright about her life. Like most of the girls in her class at school, she had had a boyfriend from the time she was sixteen. She chose him because he was the most intelligent boy in her group. They went off to joyful, State-organised camps together for the long, warm summers.

One year, 'M's cousin from California had visited between high-school and university. It was not uncommon for cousins in the West to marry cousins in the East so that the latter might obtain a visa and leave the country. The best time for this kind of liaison to occur was after graduating from high school because the exit visa did not come with a large financial penalty. However, if the individual leaving the country was a university graduate, then they might have to repay the State for the entire cost of their education before an exit permit would be issued. 'M' never said that it was the prospect of marriage that brought her cousin to visit after she graduated from high school. Romanians were often intentionally vague, crediting the listener with the capacity to fill in the gaps. Perhaps the parents may have strategized along those lines, perhaps not. 'M's fascination with meeting the first American youth, full of stories of their shared American relatives and the Californian scene might have been hard to resist. However, in this case, no marriage was contracted, no exit visa applied for. Now 'M' was a brilliant student in University with no regrets, only pleasant memories and a bright future, albeit within the confines of totalitarian Romania.

The leaves began to fall. Late autumn blooms lost their colour. Parks lost their appeal for most though not for us. The first snow turned Herăstrău Park into a wonderland for 'M' and me.

One bright winter day, 'M' had raised the hood of her coat, her only one and slightly too small for her. I loved her eager face framed in green corduroy lined with white felt that created a halo about her head. Her cheeks were aglow from the cold and the sheer joy of youth. She smiled with happiness at the beauty of the first, pristine snow to fall in Bucharest and raised her eyes to mine. We kissed.

And quite naturally, in comfortable silence, we took the tramcar all the way back to my apartment.

Cause for Apprehension?

One day after my afternoon class, 'M' and I, at separate ends of the trolleybus, were heading, so I thought, to my apartment. I was about to get off at the usual stop where the Bălcescu and Mărășești boulevards intersect. 'M' caught my eye and gave an almost imperceptible shake of her head; neither I nor she should alight. Plans changed as a matter of course in Romania, and were, I was by now fully aware, even more likely to change without explanation to protect an affair. So, we continued to ride the trolleybus several more stops. I assumed that 'M' had spotted someone who knew her, wished to avoid them, and that we would simply alter our plans and visit Parcul Tineretului – another of Bucharest's many beautiful, quiet parks where we occasionally strolled.

'M' alighted just before the boulevard took a sweep to the right and the city began to give way to countryside. I alighted from the door at the opposite end. But 'M' didn't, as I expected, head for the entrance to Tineretului Park. Instead, she entered a small, lawned area behind black wrought-iron railings that separated a Doric-looking stone building from the boulevard. The building, I reckoned, must back onto the park itself. On my solitary walks, I had passed this peaceful-looking corner and puzzled over what it might be, but I'd never dared to enter.

The boulevard was almost empty. I had closed the distance between us to 20 paces. 'M' walked confidently through the open wrought-iron gates, made for the entrance to the building, mounted the steps and disappeared inside. I sped up, slightly nervous since I had no idea where I was being led. I trusted 'M' more than I trusted anybody in Romania. I knew she wished me no harm. What concerned me was what a person might do if they were subject to intolerable pressure. At the great doors, I passed a guard in uniform. Without pausing, I bade him a curt "Good day, Comrade!" as I assumed 'M' had done seconds earlier.

When, after school hours, I worked as a teenage delivery boy for a pharmacist in Dundee, I had learned that a confident nod that suggested you knew where you were going, could get you into most buildings without being challenged. The guard said something as I passed but I made a beeline for the steps down which I'd watched 'M' disappear.

As I descended the circular staircase it became darker and I ended up in a poorly lit basement corridor with no daylight and no exit. I remembered something about the Lubyanka being an underground prison.

I was relieved, somewhat, to see 'M' waiting for me. Her smile lit up the darkness and put my mind as much at rest as I could allow it.

"Do you know where you are?" She asked.

"I have no idea!" She saw the unease in my face.

"Look!" She pointed to the stone walls. They were divided up into what appeared to be uniform blocks with writing on them. "You see now?"

I shook my head.

'M' went reverently over to one of the blocks. She touched the carved letters.

"Here is my father. In the crypt reserved for Hungarian Jews." Quietly, she read me her father's name, the date of his birth and of his early death. He had been in his early 50s. "We're in the Cenuşa, the columbarium where urns containing ashes of the cremated are placed behind remembrance plaques."

My earlier apprehension was replaced by guilt. Even a Westerner such as myself, I reflected, was vulnerable to the seeping fear that this totalitarian state intentionally induced. Ceauşescu promoted fear for his brutal party's opprobrious ends. His poison trickled everywhere, contaminating and fraying society.

31

SKIING AT SINAIA

When the Christmas holidays arrived, I let 'M' know that I was going to Sinaia to ski. I'd booked myself a room in the government villa I often used as a base for hiking. She and her mother also planned to stay in Sinaia at a villa that 'D' had access to through her position with the State Radio.

"We'll arrive on Tuesday. Let's arrange to bump into each other close to the ski-lift but accidentally." 'M' was excited at the prospect.

Tuesday evening arrived. I walked to the ski-lift. The sharp air promised fresh snow. 'M' and 'D' were three hundred metres away, daughter and mother, arm in arm. I exercised the usual caution, no acknowledgement on my part until they greeted me.

Unhurriedly, they approached, leaning intimately into each other, sharing their warmth, protecting each other against the sub-zero temperature and all other dangers. Now, I could recognise details, coats, woollen scarves wound round heads, cheeks red, breath smoking as they talked, laughed, hugged. Now I could see eyes. 'D's alert but relishing the moment, perhaps reliving happier, easier days. 'M's eyes now, smiling directly into mine. No gestures, no voices raised in greeting, only eyes, breath smoky white.

As we approached towards one another through the creaking snow,

I strained to detect the slightest message in 'M's eyes. Must I pass unrecognised and unrecognising with no more than the nod that people who pass in isolated places offer to allay one another's fears?

Yes! A warning hovered, but it was other than alarm. Perhaps unconsciously, 'M's unblinking smile was cautioning me.

"See!" It seemed to say. "Though I must adapt to my environment I don't dissemble. Here I am, 'M' and 'D', both of me! Look on us! We are intertwined, interdependent, wholly protective, with a single-double consciousness finely tempered on the anvil of necessity and pragmatism; forged for survival. You do well to be cautious; you with your single, childlike, unproven mind. This is me, us, how we are forced to be. Pay heed, tune your unaccustomed senses to the unseen, the unexpected, the un-bargained-for."

So much can be transmitted in a look; so much understood, or not.

And then the feigned surprise "Ce surpriză plăcută! "What a lovely surprise meeting you here!" Warm lips on cold hands, on cheeks. Proudly, I walked, a woman on each arm as darkness fell and sought where we might have a glass of mulled wine. Later, I walked 'M' and 'D' to their villa, then back to my own, struggling to put into words 'M's profoundly unspoken message.

A Very Special Villa

In Sinaia, my solitude had come to the notice of the maid who cleaned my room in the villa. Always a morning person, I went for walks at dawn but tended to return to my room after dinner to read and sleep early. Another reason for my isolation was that as soon as Romanians recognised I came from the West, they were inhibited. I felt their discomfort and knew the risk they ran. By retiring early, their interests, if not mine, well served.

"There are many parties in the evenings," the maid told me "You should go out. Join the fun."

The following evening, she knocked at my door. "You are invited to a party in that villa." She pointed to where, every night, I heard

The Kilt Behind the Curtain

popular music and much merriment. "I also clean there. I spoke to them. They want you to visit."

"I am from the West. I'm British. That could be a problem."

"It's no problem. Not for the couple who have *this* villa. They want to meet you."

I was surprised and curious. Everybody wanted to meet a Westerner but few dared. And so, that evening I returned from the slopes, showered and changed then made my way to the villa with the lights and the music.

Other guests, couples mostly, were confidently mounting the wooden steps arm-in-arm. I was greeted by the maid in a crisp uniform. Smiling, she escorted me to meet my hosts.

"We heard about the solitary Englishman! We wanted to meet you." I was in the company of the kind of Romanians I had never met before. They introduced themselves offering both their given and their surnames. Sică and Anica Stevoiu, both in their late 20s, greeted with warmth, curiosity and confidence.

Sică and Anica scrutinised me but not in the way I was accustomed to, as if assessing how I could be of use to them. On the contrary, they appeared to offer uncomplicated hospitality, introductions to their friends, to make me feel at home. They had no reservations about my being British, about our speaking English together in front of their guests. There was ample food and drink. In my villa, meals were scanty.

Anica wanted me to herself and drew me aside into a corner. She asked me lots of personal questions in a friendly, honestly curious way. The incredulous expression on her lovely face suggested that she couldn't imagine me coming to Sinaia and spending the evenings alone in my room. Sinaia was for mixing, dancing, laughing, having fun, a fleeting affair.

By now, I'd come to trust my intuitions and decided that Anica was neither Secret Police nor an informer. I had, I was certain, spotted Securitate agents in the room. Low-key, they merely observed, making no attempt to mix. Whereas the guests were dressed casually, the agents kept their jackets on to conceal their weapons. They padded unobtrusively between front and back doors, up and down the staircase, without mixing, occasionally exchanging looks or low whispers.

I had a feeling similar to that I'd experienced when 'M' had taken me to the crypt of the Columbarium and before I understood where I was. In this villa, I was out of my depth with few firm points of reference. There were things I recognized but I was unable to fit them together to provide a complete picture. Key elements were missing. I knew better than to fill in the blanks with guesswork.

Like most Romanians I had met, Anica was highly tuned and acutely intuitive. She watched my eyes on the Securitate agents. She sensed my bewilderment. Standing disconcertingly close to me, she explained.

"My father was director of the Securitate, the national security force."

I strove to display no overt reaction. Since arriving in Romania, I'd done my best to obstruct and frustrate the Securitate and now, here I was, in a beautiful villa in the Prahova Valley with my beloved Carpathians rising above us, in the company of the daughter of one of the ex-directors of the detested Secret Police. I was also aware that I was on the point of a helpless infatuation.

"My father died," She paused. "But our family continue to enjoy privileges."

Some of these I could see. I would find out there were many, many more.

When, a few days later, I told Anica I was returning to Bucharest, she insisted I not take the train.

"Our driver has nothing to do. I will have him take you back!" She refused my protestation.

That afternoon I found myself speeding across the Wallachian Plain in the plush rear seat of a luxurious black "Chaika". The Chaika, a Soviet-made saloon, was the preferred limousine of Khrushchev. He liked it so much that he gifted one to Fidel Castro. The Chaika was also the obligatory car for the Soviet ambassador to Romania. Anica's Chaika had leather-upholstery, off-white curtains, fold-down tables and reading-lights. It came with a driver who addressed me respectfully as "Comrade"!

A chauffeur-driven, Russian-built Chaika with curtains for privacy, was the car preferred by Romania's communist elite. Cars were few in the '60s.

32

SCHOLARSHIPS TO BRITAIN

A Delicate Task

Tony Mann, the British Cultural Attaché, invited me to a meeting. He came straight to the point.

"Romania and the UK have ratified a new cultural agreement. It allows for one or even two Romanian undergraduates to complete their degree in English at a British University, all expenses paid. This will take effect in October 1969." I thought this good news indeed! "I want you, Ron, to draw up a short list of your best students. I'll forward your list to the Foreign Office. They will make the final decision. In addition to their being outstanding students they must be able to handle life in Britain successfully."

He waited for my reaction. A sudden thought occurred to me. My drawing up such a list could present me with a dilemma. Tony misunderstood my hesitation.

"Don't you think this is a great opportunity?"

"I may have a conflict of interests, Tony,"

"How so?" Tony and I appreciated and trusted each other. I was going to have to tell him.

"I'm having an affair with one of my best students." I waited for his judgement.

Tony looked at me and then back at the papers on his desk. He pondered.

"Look Ron, you handle life in Romania very well. You're well respected both by your colleagues in the University and by the diplomats serving in our Embassy. Members of the US Embassy also respect you."

I felt flattered. It was something to gain admiration from diplomats for how I handled life in Romania.

"Be totally objective," he went on. "I can help you draw up the criteria candidates must meet. Then you draw up a list of your top students. Obtain the approval to be nominated of at least three. Give me the list unranked. The final decision is entirely up to the committee. They will have additional information obtained from other sources. All right?"

"Thanks for the vote of confidence, Tony." His trust gratified me.

"Is your friend an informant for the Secret Police?" Tony knew all about the earlier Karen business.

"In Romania, you can never be certain about anything, Tony," I answered truthfully. "I've asked myself that question many times. Based on my observations, I'd say that she's not an informant."

Although I didn't mention this to Tony, I was under the impression that this second year in-country, I had been unable to identify who my informant was. Perhaps nobody? Had the Communist Party learned enough about me from Karen and other sources that they'd decided that I posed no threat?

"Good," he said. "Because if you thought she were, we would automatically eliminate her. We can't help the Secret Police to place agents in Britain."

"I trust your judgement," Tony said and added, "Don't tell me her name. It's better I don't know."

On my way out, I exchanged news about my recent hiking activities with the wind-burned fisherman Domnul Zamfirescu, then a few words of banter with the British guards, who were cleaning their shotguns.

Tony and I drew up criteria to determine the suitability of students to study in the UK. Using them, I observed my students with even greater care. The criteria included English proficiency but also the personal and the social skills necessary for a young person to live and study alone in the UK. The successful Romanian would have to compete successfully with British students.

I said nothing at all to 'M' about the scholarships.

Finally, I'd identified four excellent candidates. Now I had the task of approaching each, separately and in confidence, tell them about their candidature for the scholarship, ask them to discuss it with their parents and approval my passing their names to the Cultural Attaché at the British Embassy. The task presented logistical difficulties. Normally, I had no contact with my students outside of class. One-evening at Capşa's with Doina, the hot-toddy with Mihaela and my affair with 'M' were the only exceptions.

Astrid

I approached Astrid first. Astrid was a reserved, modest young woman, who made thoughtful comments, had near-perfect command of English and, I thought, possessed the inner resources to succeed in a British university.

By chance, when I turned up early for class, Astrid was sitting alone in my seminar room. I greeted her and told her I had something serious to talk about. I began to explain the scholarship.

"Please do not talk to me here!" She was agitated. "Meet me at the National Museum of Art after class. Look for the room with the Nicolae Grigorescu paintings."

After my class and the goodbyes were over, I packed my briefcase and left the building. Instead of turning left, as I usually did, onto

Magheru, I crossed the boulevard with the help of an efficient traffic policeman. I always found him highly entertaining. He directed the traffic as if he were conducting the George Enescu Philharmonic Orchestra.

The Romanian National Museum of Art had been a royal palace before World War II. Its grandeur was impressive. I was familiar with its Brâncuşi sculptures and its beautiful paintings.

As I walked to the art gallery, I thought. "How can I best present this proposal to Astrid in terms that make it clear I don't have the final word?" I found her already in the Nicolae Grigorescu gallery.

We began to walk and talk quietly looking only at the paintings. I outlined the nature of the scholarship; that if she agreed, she could be added to a short list and that a committee, not me, would make the final choice. She let me finish and then looked at me very seriously.

"Professor Mackay, I appreciate your considering me for such an honour." The sad look in her eyes told me she was going to refuse but I couldn't imagine why. "I cannot let you put my name forward. My national identity card contains information that would exclude me." She saw my puzzlement and drew out her card. It bore a head-and-shoulders photo of her, her full name and additional information along with numbers and codes.

"This card shows that I'm the daughter of a Hungarian. In addition to belonging to an ethnic minority, my father had a small factory and employed people before the War. In the eyes of the Communist Party that makes him a hated capitalist. All this is recorded on my card. I count myself fortunate to have been admitted to the University. I will strive to graduate with high marks so that I may be assigned a good job. It would be unwise of me, however, to attract attention by being added to the scholarship shortlist."

I looked at her – all 19 years of this young woman – and realised that behind every innocent face of every student, lay a heart-breaking history of hurt, hardship, pain and suffering that I could know nothing about.

As so often happened in Romania, I had difficulty knowing what to say. I said nothing. I hadn't been aware that there existed institutionalised discrimination against the minorities that made up the Socialist Republic of Romania. For a brief moment I imagined what 'M's reaction might be when finally, I approached her about the scholarship.

Astrid and I continued to stroll from hall to hall looking at paintings, saying little. We entered an empty room from one direction and Petru, the older male student in my class, entered from the other. He came to greet us as if the encounter were a coincidence. As soon as he appeared, I remembered having glimpsed him earlier as I walked from Pitar Moş to the art gallery. Suddenly, images of having sighted him in other unexpected places on odd occasions flashed into my mind. I'd felt a little sorry for Petru, believing that he must feel like a fish out of water among the younger students in the class and so I did my best to be kind to him and make time to speak to him whenever he initiated contact.

Petru joined us. Astrid looked distressed. She made a quick escape and left me stuck with Petru.

Astrid's revelations had left me in an uncongenial frame of mind.

Petru appeared unfazed by his fellow-student's hurried departure. Now he was at my elbow, asking tedious questions as I made for the exit. We stepped into the sunshine and I turned on him.

Petru's Secret

"Petru, you followed me from class to the art gallery!"

He remained silent but his face told me I was right.

"Petru, I give you a generous amount of my time both in class and out." He nodded. "I really don't appreciate your following me or a fellow-student." The words erupted with more anger than I'd intended.

Petru looked as though he might burst into tears. That made me all the more annoyed. Why was I taking my frustration out on this poor man?

"Professor Mackay," he regarded me sorrowfully. "Please let me offer you my sincere apologies!"

My trolleybus was approaching. I waved his apologies away.

"Please!" he begged, "Let me explain." He seemed to be suffering so I let my bus leave. "I am engaged by the Securitate to inform on you this year." Petru now had my full attention. "I completed my first degree in a different department several years ago. Now this is what they make me do. It's my job."

I was taken aback. I hadn't suspected that Petru was my informer. I wanted to know more.

"How can they make you do something like this? Can't you find a job you like and forget about Securitate?" I could pose a simple question but for a Romanian life under communism was not simple.

And right there, in the winter sunshine on Magheru Boulevard, with people walking by in both directions and trolleybuses showering sparks, Petru outlined his story. Six years earlier he had been picked up by two Securitate officers and driven to their headquarters. There, they confronted him with undeniable evidence of his homosexual liaisons.

"'You will be prosecuted and jailed,' they told me, 'unless you agree to cooperate fully with us.' What option did I have?"

And so, for years, he was assigned to get close to and inform on visitors from both West and East.

"Why?" I asked.

"The Communist Party feeds on suspicion. Especially of foreigners and especially those from the West."

"What does the Party suspect them of?"

"Anti-revolutionary activities." I suspected that phrase meant as little to him as it did to me.

He went on. "I am assigned principally to inform on visitors who might be homosexual and to compromise them if possible."

I listened, as much heartbroken for this poor, unhappy young man as appalled at the Romanian Secret Police. I also felt offended at the implications of what Petru was saying. Why would the Securitate think I was homosexual? And then I remembered Karen and how I had rebuffed her advances the previous year. Now I understood! In our drama course at Aberdeen University we had read William Congreve's tragedy The Mourning Bride. His classic line came to mind.

"Heaven has no rage like love to hatred turned, / Nor hell a fury like a woman scorned."

"I'm very sorry, Petru!" And I was. I was sorry for him, for his entrapment, for the blackmail he was suffering, for his being prevented from getting on with life like any other normal human being. And I felt just as sorry that I had treated him with unkindness and anger after Astrid left us. "Very, very sorry, Petru," conscious that I was repeating myself but quite unable to find anything more adequate to say.

My apology, my commiseration with his condition, seemed to cheer him. He almost smiled. "It's all right, Professor Mackay, I will report to Securitate that you became annoyed with my persistent attention and that you asked me to keep away from you. I will give my report in such a way that it will harm neither you nor me. They will listen, accept, and give me a new assignment."

I wanted to weep at this poor man's unresisting acceptance of the lot he had been cast, of Securitate's cruel stranglehold over him.

"But before I offer Securitate my report, Professor Mackay, let me invite you to my home. I would like you to meet my parents and my sister. I would like to help you understand our country, its legacy."

He was concerned for *me*, wanted to offer me an insight into how Securitate could degrade and brutalise its own citizens.

I agreed that I would accompany him to his home for *masa de prânz,* lunch, two days hence. And so, two days later we met on Magheru and walked together to where he lived with his mother, father and sister not far from Piața Romana.

33

MUTTON CHOPS WITH PETRU

Petru, his aging parents and his younger sister lived together in a single cluttered room in what had once been an elegant single-family home. Other families lived in the remaining rooms. They shared the bathroom and kitchen. There was evidence that Petru had rearranged the furniture to make room to welcome me for lunch. Curtains that afford individuals some privacy, had been drawn back

I felt as though I were in an auction saleroom where sundry items had been hastily stacked to make way for customers to follow the auctioneer and offer bids on random odds and ends.

Petru's aged parents greeted with humility and old-fashioned grace that intensified my discomfort. I was the cause of Petru's failing his current mission, the occasion of this invasion of their privacy. Now I was a witness to their humiliation. "I didn't create the system," I reminded myself. "The blame lies with Marx, Lenin, Stalin, and now with the maniacal dictator Ceaușescu."

I hoped that the Securitate would accept Petru's explanation for ending his surveillance of me and offer him a more congenial assignment. But then I thought, "No! Petru should never have been entrapped into the degrading mire of eavesdropping and informing in the first place! The system has no right to treat a human being this way!" I raged at my inability to do nothing but stand helplessly by and bear witness.

His parents were in their fifties but looked decades older, faces worn and lined, once-smart clothes threadbare. His sister greeted me "Good day" then not another word. She was probably mortified that the first guest they had had in years had been invited for the sole purpose of witnessing her family's degradation.

Petru offered me a matter of fact commentary on the scene as if explaining exhibits in a museum to a foreign visitor.

"This is how we live. How we've lived for a long time now."

His father and mother offered apologetic grimaces, nodding the truth of Petru's toneless remarks.

"Before the War, my father had had his own dental practice. My mother was an opera singer. This entire house was our home, inherited from my grandparents. After the War he was permitted to work as a dentist, but his practice was nationalised. He became a State employee."

"With many houses damaged during the War and large numbers of displaced Romanians arriving in the city, the government nationalised all private homes. The State allocated living space on a square-metre per person basis" as they had done, I remembered, in 'M' and 'D's home. "My father and mother had the good fortune to retain one of the larger rooms. We share the kitchen and the bathroom with other state-appointed residents."

There had been little work for an opera singer. His mother supplemented the family's income by giving voice lessons. For some reason that Petru didn't go into, his father had blotted his copybook with the Communist Party and as a consequence was forbidden to continue practicing as a dentist.

"Now he works as caretaker and cleaner at the dental practice he himself established before the War. Every morning he opens to door to his former colleagues. In the evening, he sweeps up."

Petru's mother and father sat nodding cheerlessly as if listening to a casual story about the misfortunes of distant acquaintances. While Petru talked, his sister placed a large plate at each setting. In the middle of each plate was a huge mound of mashed potatoes. The centre of each had been scooped out and filled with hot dripping. In the dripping swam a large mutton chop. I ate it with relish. Mutton was a staple in Scotland. Indeed, I never heard of anybody eating *lamb* until I was an adult.

Petru had wanted me to understand his situation; now I did. I added inhumanity, cruelty, malevolence and vindictiveness to the atrocities of the Romanian Communist Party and its Secret Police.

A village house in Transylvania, home to many Hungarians.

34

SCHOLARSHIP CANDIDATES

Astrid had surprised me by turning her candidature down. I approached my next two candidates individually, warning each to treat the matter of their nomination in absolute confidence. It would not be appropriate for anyone to know whose names I would put forward. Fortunately, Romanians were used to keeping secrets. Every Romanian family had confidences that they could not afford to have leaked without causing extreme risk. Children, from just a few years old, were coached to learn by heart the "authorised" version of their family history for use in school and in all public institutions.

I was heartened when candidates two and three, a young man and a young woman, separately listened to my explanation of what was involved in agreeing to be nominated. Each made a best effort to conceal their excitement. Each went home, discussed the matter with their parents then met me once more to approve my putting their name forward to our Cultural Attaché.

Now it was time to talk to 'M'. Of course, she would not learn the names of the other candidates.

Astrid had established, in my mind, the Nicolae Grigorescu room in the art gallery as the ideal neutral location for talking about the scholarships and it was there that I had met students two and three. So too, it was before the Grigorescu paintings that I told 'M' that the British Government would cover the entire cost involved in a Romanian student completing their undergraduate degree in a British university.

'M's naturally happy face lit up with even more delight. "You must have 'D's permission before I can add your name to the short list." Although I didn't tell her whose names were on it, she must have had an idea. Students were highly competitive and knew their approximate ranking in the class.

"The successful candidate must him or herself, apply for and obtain, a passport and exit visa." 'M' bounded off to recount all to 'D'.

The following day we met once more in the art gallery taking the usual precautions. 'M' was as excited as if she had already won the nomination.

"Remember that a committee will make the final selection. The committee's nomination will mean only that the offer has been made. It's up to the nominated student to undertake all the paperwork and clear all the hurdles it takes to be awarded an exit visa."

I needn't have been so didactic. As usual, 'M' was steps ahead of me. "We talked about it. We understand." And as if to bring *me* back to reality, added, "If I'm selected, we will take care of the rest." She seemed confident. So young, yet with such a profound grasp of life.

From the look on 'M's face I could see that she was aware she was about to begin an epic journey. For her, the University, Bucharest, her friends, our affair, the Socialist Republic of Romania, absolutely everything, with the sole exception of her mother 'D', were milestones on that journey. Some of these milestones would be welcomed when reached, some might mean more than others, a few would be remembered and fondly missed, but all milestones would be reached and allowed to recede into the distance. 'D' was the one, the only one, who would not be waved *"Goodbye!"* on 'M's pilgrimage. This insight came to me as something I knew to be true, and that truth liberated me.

Of course, 'M' and 'D' had discussed it! *Of course*, they knew better than I what was involved in negotiating the shoals and reefs they sailed in! There was no doubt in my mind that they were, together, eminently capable of navigating the bureaucracy so that they would end up with all the paperwork, stamped and approved, everything that would be necessary to allow 'M' to leave the country. Like so many Romanians, 'M' and 'D' were not merely survivors keeping their heads above water, they were resourceful swimmers who, by virtue of their own efforts, intelligence, and canny enterprise would always thrive, progress, and prosper irrespective of any obstacles in their path. It would not be easy for them, but they would strive with such elegance that they would make their journey appear effortless.

Romanians, in my experience, sought no applause for effort; they valued recognition for achievement. Both 'M' and 'D' deserve deep respect for the victories they subsequently won.

The magnificent Dealu Monastery close to Târgoviște was built in the 15th-century and dedicated to Saint Nicholas.

35

ALEXANDRU FROM ALEXANDRIA

The People's University

In addition to my required teaching at the university, I volunteered to teach an adult education class in English literature at the "People's University" one evening a week. I found my teaching duties overly light. I also wanted to give something back to Romania for the unforgettable experiences I was getting from my stay. My evening class at the People's University met for two hours a week and consisted of six exhausted adults. Only five turned up to any given class and not all had done the weekly assigned reading so the classes tended to be informal. The students' proficiency in English was poor with the result I did most of the talking which seemed to suit them just fine.

After the very first class, one of these students, a middle-aged man named Alexandru, told me that he lived in Alexandria, a city distant from Bucharest where he managed a factory. Distance and unforeseen circumstances would keep him from attending every class. However, he did have to visit the capital twice a month on business and had a couple of hours free on these afternoons.

"Would you be kind enough to meet me twice a month to talk about the curriculum and make sure I do not fall behind?" He mentioned a well-known pastry and coffee shop.

"Of course!" My efforts to meet more Romanians were paying a dividend.

"I will call you before I come to Bucharest. We can go over the classwork, have coffee and chat."

On one occasion, 'M' suggested coming over to my apartment the following day. I apologised and explained about "Alexandru from Alexandria" making a joke of it. I'd never mentioned him to her before because I intentionally kept separate the various strands of my life in Bucharest. 'M' showed keen interest in the fact that I met this man.

"What time is your meeting with Alexandru tomorrow?"

"At two o'clock."

"Would you mind if I came into the coffee shop while you're there?"

"Why?" I thought it an odd request.

"I just want to see him."

I saw no harm in it. 'M' would enter the coffee shop about 20 minutes after my appointed time with Alexandru. Neither she nor I would acknowledge one another. "M" would buy profiterole at the pastry-counter and leave. I gave her the lei to purchase the profiterole as a present for her and 'D'.

Regularly, Romanians would make odd requests without offering any explanation. They were obsessed with security. I thought they exaggerated the capacity of human agents to be everywhere at all hours. Nevertheless, because I had time on my hands, I would sometimes take a very long trolleybus or tram ride at the very time I thought that an agent assigned to follow me was likely to be getting ready to change shifts. I laughed to think what his irate wife might have to say when he eventually arrived home hours late for his supper.

My opinion was that the Secret Police used the twin tools of fear and uncertainty to make the average Romanian believe that its network all-seeing and all-hearing. I was wrong about the capabilities and cunning of the Securitate.

"What did you think of Alexandru," I asked 'M' the day after my last meeting with him.

She smiled and shrugged. "Thank you for the profiterole. 'D' and I enjoyed them." Romanians, even as young as 'M', were skilled at changing the direction of a conversation if it suited them.

As it happens, 'M' was to meet Alexandru again, several weeks after I had completed my contract and left Romania for good. That meeting was a scheduled one-to-one interview in his office. His office was not in a factory in Alexandria but right there in the centre of Bucharest. Alexandru was dressed in the uniform of a senior Secret Police officer.

A winning smile from a resident in the Danube Delta.

36

CROSS-BORDER TRAVEL

To Sofia on the Moscow Express

"Ronald, would you be willing to travel to Sofia in Bulgaria once a month to offer two days of intensive classes to final-year students at Sofia University?"

Tony had called me to an emergency meeting in the Embassy in the last months of 1968. The visiting British professor in Sofia University was indisposed. He'd fallen ill. To save the bilateral exchange agreement between the UK and Bulgaria and buy time until a replacement could be found, Tony told me that "the powers that be suggested" I fill the gap. I took the invitation as a compliment.

"Of course, I will!" This was exactly the kind of exploit that appealed to me. What is more, my teaching schedule in Bucharest made it possible.

"The Moscow-Istanbul Express stops in both Bucharest and Sofia. From here, Sofia's and eight-hour journey. The budget won't rise to a berth, you'll have to sit up overnight there and back, but your expenses will be paid. You will have the use of the British professor's apartment in Sofia."

I was learning that the British Council, the British Government's cultural arm, tended to treat its own career officers well but its contract employees less so. But I was young, enthusiastic, and used to hardship. Sitting up overnight was a small price for the thrill of taking on this new adventure.

This additional assignment meant I'd be making four visits to Sofia before the end of the year.

When I went to board the Moscow-Istanbul Express, I was excited. The station smelled of hot oil and diesel. There was the slamming of doors, the urgent shouting of porters. Final crates were being loaded. Porters, mysterious passengers and armed Militia abounded. I was escorted to my solitary compartment by two Militiamen. "Until the train leaves the city limits, your compartment door will be locked." They saluted and left. Two bench seats faced each other upholstered in an ugly green plasticised material.

Once the train was underway, a uniformed employee tramped down the corridor unlocking compartment doors. I rose to stretch my legs and found uniformed officers at both ends of the corridor. I noticed that in each compartment there were passengers who had boarded in Bucharest, but I was the only Westerner. That neither surprised nor bothered me. By now, I was used to being on my own. What I wanted to do was to pass from my carriage into the other carriages to get a glimpse of the passengers who were travelling all the way from Moscow to Sofia but the Militiamen stopped me.

"You may use the toilets at either end of this carriage, Comrade, but you can go no further"

"Is there a dining car?"

"The dining car is not available to you. An attendant will bring you refreshments."

I was disappointment. When I'd arrived in Bucharest from Paris in 1967, I'd been too inexperienced and had insufficient money to really take advantage of the legendary train-trip. Now I was a little more worldly and had money to enjoy a meal in the dining car. But it was not to be.

Before we crossed the Danube into Bulgaria, an attendant delivered a small package of food and a soft drink to each passenger. So much for a romantic dinner in the company of mysterious travellers!

The light in my compartment was too weak to read by so I stretched out and slept until we reached Sofia before dawn. At the station, I was met by a Bulgarian driver from the British Embassy and driven to an apartment block. When the concierge was satisfied as to my identity, he escorted me to an apartment and gave me the key.

The apartment had been assigned by the Bulgarian Government to the British professor I was replacing. It was like my own in Bucharest. The apartment had shelves of the professor's books and he'd decorated the walls with pictures. In fact, he had made himself quite comfortable.

I slept for an hour and was jolted awake by the telephone.

"I'm a secretary from the British Embassy. I'm waiting downstairs with a car to take us to the University."

She was young, English, and friendly. She suggested we stop for a coffee and pastry on our way since I'd had no breakfast. It had snowed overnight. The city was white, pristine, and beautiful under a clear blue sky and a bright winter sun. There was little traffic. The boulevards were wide, the buildings magnificent, the pedestrians dressed in winter coats with astrakhan collars and hats as in Romania.

"This is one of the main sights." The car was passing the most ornate Orthodox cathedral I'd ever seen. "The Alexander Nevsky Cathedral."

I was awe-struck by its massiveness and its golden domes that reflected the sun as if offering me a personal welcome to the city. Already, I loved Sofia!

We had breakfast in a crowded coffee shop. Customers came in out of the snow with wide smiles and cheeks rosy from the cold. They peeled off scarves and layers of clothing and greeted one another as if they hadn't a care in the world. "Are there places like this in Bucharest," I wondered, "where young people hung out together?" If there were, I hadn't found them or hadn't looked hard enough. In

Bucharest I was too aware that my presence embarrassed people. In Sofia I was a transient and so I cared less.

"Ready to meet your students?" We re-wrapped ourselves in coats, hats and scarves and drove to Sofia University. She took me to the Faculty Room. It was empty.

"Not to worry, you'll meet them later," She walked me up-stairs. Students and faculty were rushing to and from class. We reached an open door. In the tutorial room, twenty-four pairs of eyes studied me intensely. The all seemed to belong to attractive young women, slightly older and more sophisticated than my students in Bucharest. They were all in their final year of studies.

The secretary, with whom they appeared to be on familiar terms, introduced me and explained that I would replace their regular British professor. We would have two seven-hour days of intensive classes each month together. To my embarrassment they applauded.

"I'll come back for you at noon. I've arranged for us to eat in the faculty dining room so you can meet the head of the English department and some of your colleagues." The secretary left with a casual wave that the students cheerfully returned.

Together, my students and I spent the rest of the morning working out exactly how best I could help them for the two days each month that I would visit. The students were business-like and clear about what they wanted from me. They demanded lots of interaction in English so that they could practice with a native speaker. Fortunately, my Scots accent didn't bother them as it had Madame Cartianu. They wanted constant correction and feedback from me. Finally, they wanted me to address issues related to phonology, morphology, syntax, sentence structure, vocabulary, idioms, and irregular and phrasal verbs.

This kind of participatory planning with willing undergraduates appealed to me. My new Bulgarian students were just as diligent and as willing as the Romanian students I was so proud. We worked so effectively that before we broke for lunch, we had a solid workplan for the rest of my stay and a good idea of what we would do when I visited again in November and December.

As my students left for their dining hall, the secretary returned and escorted me outside into the campus, now white with snow and bathed in warm winter sunlight.

The faculty dining room, just a short walk away, was a grand place with velvet curtains and painted plaster frieze-work that warmed the heart. A huge table had been laid for us bright with linen tablecloth and crisp napkins, polished cutlery, and sparking glasses. Most members of the English department were already seated and rose to greet us, the secretary first, she was clearly popular with them, and then me. All were affable. The chairman thanked me to agreeing to rescue them from this emergency.

To return to my students, I had to leave immediately lunch was over. The secretary told me that she'd drive me to my apartment later. I felt coddled. The departmental chairman shook my hand and so did each of the score of faculty members. As I expected, I never saw any of them again.

My students and I did a solid afternoon's work with many a laugh shared, and when I left the building, the secretary was waiting for me. As she dropped me off at my apartment building, she said, "Now you know everybody, I'll send a driver to take you to and from the University tomorrow. He'll also take you to the train station in the evening in time to catch your train to Bucharest." I appreciated the attention.

The following day my students and I again applied ourselves diligently to get through our work plan and we succeeded. I was driven home. The concierge handed me an envelope. "Professor James E. Augerot will call for you at 6.30 p.m. and bring you to the American Embassy." This was odd, since I had not yet met the British Cultural Attaché.

I lay down to recover from the labours of the day and fell sound asleep. There was a knock at the door.

Jim Augerot

A tall, lanky, goatee-bearded man with astute eyes and a twisted grin thrust out his hand, "Jim Augerot! Welcome to Bulgaria, Ron!" Jim was one of these Americans who immediately made you feel embraced. He particularly wanted to meet me because he and his family had spent a year as Visiting American Professor at a University in Romania a couple of years earlier.

We discovered that we shared a common passion for the country. I felt I'd known Jim all my life. Jim was different from the visiting American professors I knew in Romania. He had a better insight into that country, a better relationship with the people and had learned fluent Romanian. As a Slavic language specialist, he also mastered Russian and Bulgarian.

Jim was currently the Visiting American Professor at Sofia University on exchange from the University of Washington. He and his family lived in the same apartment building I'd been allocated in Sofia. He took me upstairs to meet his beautiful wife and charming daughters aged seven and nine.

When their baby-sitter arrived, Jim and his wife kissed their little girls and we left together for the American Embassy.

The party, as well as others I would attend later, was held in the Embassy building itself. American cocktail parties tended to be livelier than their British equivalents. Americans made a point of introducing themselves and asking newcomers questions to put them at ease. Jim and his wife were obviously a popular couple and had lots of friends among the American diplomats and the visiting American graduate students. They saw to it that I was introduced to everybody. As a young Brit among all these worldly Americans, I felt like the guest of honour.

The Cultural Attaché, who looked more like a benign Mafia Don than a diplomat, engaged me in a lively conversation about my work in Bucharest as well as in Sofia. He took a real interest in me. We talked for a good fifteen minutes. I'd already learned that it was a mistake to judge Americans by their looks.

"If we can be of any assistance to you in any way, call me!" He gave me his card, shook my hand generously and moved on to talk with others.

Second Visit to Sofia

The following month I made my second trip to Sofia. The intensive workplan that my students and I had established paid off. We made great progress much to their and my satisfaction. As we broke for lunch, a middle-aged woman entered the room.

"Professor Mackay, I have been waiting for you to finish your class. I am Radka Draganova, coordinator of English in Sofia's technical schools. I need your advice." It never ceased to amaze me how enthusiastic and dedicated English teachers could be and I always did my best to help.

"I am trying," Madame Draganova told me, "to tailor instruction to the precise needs for which our students will eventually have to use their English."

I'd become interested in English used for professional, scientific and technical purposes when I taught in Bournemouth the previous summer, so I was sensitive to her challenge.

We had a hurried lunch together talking about the difficulties of designing and teaching classes to students whose use for English was primarily related to their work. Such students were often impatient of general courses that dealt the trivial daily lives of imaginary characters. By the end of lunch, I'd agreed to find time to visit her technical schools to become familiar with the kind of students she was dealing with. I promised to talk to her teachers in an effort to win them over to a more pragmatic "English for special purposes" approach.

Fortunately, I was staying in Sofia longer on this Trip. Jim Augerot had persuaded me to stay over for a few days and spend the weekend at a villa with his family and some close friends who worked for the State Department.

Madame Draganova turned out to be a very dynamic and visionary teacher of English. She'd been grooming a cadre of good teachers to buy into her 'Technical Purposes' approach to teaching English. This, she believed, would both motivate technical students and provide them with English to support their work skills.

"Is there any way that the British Embassy in Sofia might assist in any way? Technical English dictionaries, technical reading material, visitors – anything to help my teachers?"

It never ceased to humble me how enthusiastic teachers like Radka could be about teaching English, the language of a country they'd never visited and seldom heard spoken by native speakers. I remembered the words of "Mr. Jones", the Head of Mission with whom I'd enjoyed the carbonized crêpes at the Athénée Palace in Bucharest the previous year:

> *"The Foreign Office considers English teaching to be the sharp end of our trade initiative with Central and Eastern European countries, Mr. Mackay. We see your role as preparing the ground, so to speak!"*

Optimistic, I promised Radka that I would speak to the Cultural Attaché at the British Embassy and find out what help they might offer.

I had not met the British Cultural Attaché in Sofia. When I returned to Bucharest, I asked Tony Mann to arrange a meeting for me on my next trip to Sofia. Tony, as ever, was cooperative. He told me that he had no real counterpart in Sofia, no British Council officer in disguise as a diplomat, but he promised to do his best.

Third Visit to Sofia

The following month, Pearl was coming back to Romania for several weeks. Because she was changing jobs, from Queens Gate Girls' School to the Medical School at London University, she had free time. I knew she'd be delighted to visit Bulgaria as well as see more of beautiful Romania.

On my next trip to Sofia, Pearl accompanied me. The British Consul, Doris Cole, lent me her car and we drove south-west towards Giurgiu to cross the Danube using "Friendship Bridge" that linked Romania with Bulgaria. It was winter and road conditions were poor. However, the sun was shining brightly when we left Bucharest and I'd learned to drive comfortably in snow, so the journey didn't cause us concern despite the fact that we passed an alarming number of tractor-trailers, that had jack-knifed on the ice.

There wasn't a single car when we arrived at the heavily-guarded border and so we suffered no more than the ordinary delay involved in passport checks, obtaining tourist visas, changing the obligatory sum from US dollars into Bulgarian Levs and having the car thoroughly inspected. Inspections included the use of sniffer dogs and mirrors on wheels inserted under the chassis of the car. The border guards, in bulky topcoats, huge felt boots over their leather uniform boots and handsome trapper-style sheepskin hats to warm their heads and ears eventually re-shouldered their rifles and machine guns and waved us impassively into Bulgaria.

I'd told my students at Sofia University that my mother would be coming with me and several had offered their services as guides. That surprised me; my Romanian students never dared to make such offers.

At the appropriate hour, I presented myself at the British Embassy for my appointment with the Cultural Attaché. I was asked to wait in an imposing anteroom for a good 20 minutes. Finally, the Attaché's secretary escorted me into his office. He stood and shook my hand in a way that suggested my visit was not the highlight of his day.

"You asked to meet me? What about?" I was taken aback. He seemed to know nothing of my request.

"Your counterpart at the Embassy in Bucharest made the appointment." My reminder failed to jog his memory. "About possible assistance to the Technical Schools in Sofia?"

"Ah!" He recollected, "Mr. Mann of the..." he paused, "...the British Council." Ever so subtly, he was telling me that Tony Mann was only an employee of the British Council, that Tony's appointment as a Cultural Attaché was a temporary one determined by convenience while he himself was a career officer in the British Diplomatic Service. Some members of the Diplomatic Service guarded their status jealously. They could resent those from other services such as the military or the British Council whose diplomatic status was conferred temporarily.

I was 26 years old and the gentleman across the desk from me was in his 30s. I was earning £30 Sterling a month. He must be earning ten times that plus innumerable perks. Both our remunerations came from British taxpayers. I had volunteered to undertake additional tasks in Sofia once a month, on my own time, and free, to help resolve a problem that fell under his jurisdiction.

Swallowing my annoyance, I explained the background. How I'd been approached by Madame Draganova; her enthusiastic work as English language coordinator in Sofia's technical schools; her innovative ideas about teaching English for technical purposes; the need to motivate her teachers; the dearth of materials and training.

When I'd finished, the Cultural Attaché gave me a bored look. "So, what do you want?"

My heart lightened. He might lack social grace but at least he had captured the crux of the matter.

"You might donate technical dictionaries, trade magazines; commercial brochures that advertise and explain technology. You might search for a way to send Madame Draganova to the UK for a short training course on how to teach technical English and develop suitable teaching materials."

My suggestions kindled no enthusiasm in the Cultural Attaché. Noticing that he hadn't taken a single note, I added, "I could make a written list of recommendations if you like."

He extracted himself unwillingly from his comfortable chair, went over to a small coffee table, picked up a glossy, illustrated magazine and tossed it to me. I caught it. "Country Life" with a portrait of a woman in riding dress on the cover and the promise inside of articles on shooting and fishing on rural estates.

"I might be able to rustle up past issues of these,"

"These are not suitable." I dropped the glossy Country Life on his desk. I knew I was burning my bridges. "Thank you for agreeing to see me, Sir."

Visibly relieved, he escorted me to the door. I passed into the refreshingly cold sunshine and snow.

That same week Jim Augerot and his wife engineered an invitation to a reception in the American Embassy for Pearl and me. There were lots of happy people, the noise of animated conversations, and liveried servants circulated discreetly with trays of drinks. Pearl was spirited off by someone interesting in order to meet someone even more interesting. She could hold her own in any company.

The American Cultural Attaché who looked like a Mafia Don spread his arms in welcome. "Raawn! Great to see you, Raawn!" I barely recognized my name. He attempted to wring my arm out of its socket. "Ah was just talkin' to your lovely mom, Pirl! Now I know where you got your good points! You sure hit the jackpot with a mom like Pirl!"

I accepted his compliments and got right down to business. "I need your help with something important!"

"Raawn, you know you just gotta tell me what it is an' I'll help!" He was sincere.

I told him about Madame Radka Draganova and her plan to teach technical English. He listened intently to my suggestions as to how we might help. He asked me some pertinent questions, nodded at my replies then called over his aide.

"Escorts, please! Now!"

Two armed Marines arrived resplendent in full uniform.

He saw the surprised look on my face. "Sorry! Regulations! You Scotties gotta be escorted in case you pocket our silverware!" He laughed as we descended in the elevator, a Marine on either side of me. "You can choose whatever you want!" Choose what, I wondered.

We exited into a room stacked with books, all new. Literature, language courses, dictionaries.

"Raawn, you just take what you want for Draganova. Tell her Compliments of the American Embassy!"

I spotted an opportunity here that could benefit both Madame Draganova and the Americans.

"What if I take just these for the moment," I said grabbing some technical dictionaries, "I'll ask her to invite you to visit her technical schools. You can assess her situation and maybe invite her to select materials for herself."

"I don't want to step on my British counterpart's turf. Brits can be–"

"Don't worry," I interrupted. Our Cultural Attaché is unable to offer any help at all."

"Then it's a deal," he signed and handed me his card. "For Madame Draganova. Ask her to invite me through Protocol. She'll understand. I'll see her right!"

Business done, with our Marine escorts we returned to the reception upstairs. Before he left me, I turned out my pockets. "See? No silverware!" We both laughed.

"If, as Mr. Jones from the British Foreign office had said in Bucharest, language is the sharp end of the battle for commercial trade, I'm happy for the Americans to win it." I told myself. "They know how to grasp the initiative."

Pearl had a wonderful time in Sofia. We managed a trip to Vitosha Mountain for the spectacular view of the elegant city of Sofia. She, I, my students, Madame Draganova and the American Cultural Attaché would all gain from my temporary appointment to Sofia University. I felt triumphant.

37

A TROOP OF TANKS

Howard's Visit

Howard Owens, a close friend, wrote saying he'd like to visit me in Romania in the spring of 1969. Howard was in his mid-twenties, affable, smart and with a desire for adventure. He worked as a commercial lawyer in the City of London. I knew him to be cool-headed, reliable and resourceful, all prerequisites for travelling in Communist Romania.

Few British people travelled behind the Iron Curtain in the 1960s. Even fewer travelled to the more distant rural areas and villages of that picturesque country sandwiched between the USSR and Bulgaria.

Howard arrived for the spring vacation. We had decided on a trip east to the Danube Delta, north to Moldavia and then south-west through Transylvania and back to Bucharest.

Once again, the generous British Consul lent me her Ford Anglia.

A week before Howard and I were due to leave, I told 'M' about the trip. A couple of days later she asked if she and 'D' could accompany us. I explained to 'M' that we'd be driving in the British Consul's car with CD plates that always drew attention and that we'd be roughing it in out-of-the-way regions.

'M' and 'D' were adamant about coming with us. I always left the final decision in these matters to the Romanians themselves. They alone knew the risks. 'M' and 'D' would make interesting companions with lots of contemporary and historic observations to make. Their presence would undoubtedly add interest.

We picked up 'M' and 'D' at the central railway station in Bucharest. They didn't want the car to appear at their home. Each carried a small travel bag. I had insisted we travel light. Settling into the back seat, 'D' was saying something important to 'M' in Hungarian.

As I turned the key in the ignition 'M' asked, "Can you show us on the map the route you plan to take?"

Whenever I was lent a car by the British Consul and planned a trip, I had to lodge an itinerary with the British guards at the Embassy. I'd done so for this planned journey and I'd already explained the itinerary to 'M'. Nevertheless, I opened the map and traced the planned route – a loop that took us east to the Danube Delta, north into Moldavia as far as Siret in northern Bukovina on the USSR border and then back in a south-westerly direction through southern Bukovina into Transylvania to Brasov, through the Prahova Valley and home to Bucharest.

When I finished, 'D' had another conversation in Hungarian with 'M' then 'M' asked, "Can we make the trip in reverse?" Apparently, this is what 'D' wanted.

Other than representing a radical departure from the itinerary I'd lodged with the Embassy; it made no difference to Howard. What I believed was of primary importance to the Embassy guards were merely our departure and return dates, so I agreed to D's request. We headed across the plain to by-pass Ploiești.

As I drove, I reflected that what had just transpired was a typical Romanian transaction. 'D' wanted to travel through Transylvania first. Her English and certainly her French were good enough for her to have asked me herself, but she had chosen to make her request through a second party, in Hungarian, a language that only they shared.

I'd observed that Romanians often made requests to a third party, usually me, through a second party. By doing so, if I denied their request, they didn't lose face. If I granted their request, they were relieved of having to express their gratitude directly to me for the favour. That meant I had no right to demand a favour from them in return. Romanians avoided incurring favours. In a society in which individuals were hypersensitive to debts incurred and balances outstanding, it was a way to side-step a debt in the first place, no matter however small. The use of a language unintelligible to me further distanced the person making the request from me who held the right to grant or deny it. Byzantine reasoning!

We by-passed Ploiești with its reeking oil refineries and stopped for lunch in the majestic Prahova Valley. 'M' and 'D' parted company with us saying they'd meet us as we readied to leave. Predeal was a popular resort, many state enterprises had villas there assigned to them by the Communist Party and I assumed that 'D', who worked for State Radio and Television, didn't want to risk being spotted with Westerners.

After lunch we continued. It was still early when we stopped in Brașov. We were uncertain whether to stay the night in Brasov or continue. 'D' resolved the matter by suggesting we spend the night there. She and 'M', she explained, would stay with friends while Howard and I would be free to do as we pleased. Brașov was an interesting city that deserved time to explore, so we left the women in the centre, promising to pick them up the following morning after breakfast at 9.30.

Howard and I went to visit the fairy-tale-like Bran Castle with its turrets and spires and bright red gables. Howard had read the novel about Count Dracula and his eerie castle in Transylvania. The author, Bram Stoker an American writer, used to spend his holidays near Montrose in Scotland. Stoker had used Ecclesgreig Castle in Kincardineshire, Scotland as his inspiration for Dracula's castle, not Bran.

When I told him, Howard accused me of enticing him to Transylvania just to boast about my native land!

The following morning, we collected 'M' and 'D' in the square dominated by the Black Church. They'd enjoyed an evening with friends. Now we were ready to head for Bistrița, 300 kilometres to the north. We planned to take it easy, stopping often to admire the beauty. There was no traffic at all, making driving a joy. Cars were generally owned by the State and used for State purposes. The road took us through the Olt Valley and then undulating, forested country with villages that had been populated now by Romanians, now by Saxons, now by Hungarians for centuries.

It was a fresh, spring morning and the world was bathed in early sunshine as we approached Sighişoara. Once more, 'D' conversed with 'M' in Hungarian. "Another request!" I thought. Sure enough, 'M' asked if we might enter the town and drive through its quaint streets. Her father and mother, she explained, had known Sighişoara well. 'D' wanted to relive happy memories. That made perfect sense.

Guided by 'D', I drove very slowly from one narrow, cobbled, traffic-free street to another. The streets were lined by yellow, red, and ochre houses separated by small, shaded, picturesque squares.

As we turned a corner 'D' gasped.

"Stop the car!" I did. "Please return to the main square. We will join you there within the hour." The two women got out of the car and hurried off down a narrow side-street.

Howard and I parked in the centre of this fortified medieval town and walked around admiring the buildings and people going about their daily lives.

"This could be Reigate in Surrey!" Howard was referring to the apparent normality of everyday life in this ancient town. I was glad that Howard was seeing the normal life of Romania. Even in a totalitarian regime imposed by the Romanian Communist Party and its apparatus of repression, it was important to see that people still lived their lives with a semblance of normality.

Shortly, 'M' and 'D' returned in the company of two well-dressed gentleman in their sixties. 'D' introduced them as 'old friends'. I invited them to join us for coffee and a pastry. Both gentlemen, one with a shock of pure white hair, the other with thinning dark hair, scrutinised Howard and me as only Romanians can. I was used to close inspection and had forewarned Howard. He played along beautifully with his usual friendly smile. The gentlemen seemed satisfied. They explained that a town festival was due to start. Perhaps we would like to watch it? 'M' and 'D' looked at us expectantly. We agreed.

I suspected that 'D' had set up this "accidental" meeting before she left Bucharest. This, I'd learned, was how Romanians did things. It saved offering explanations. It prevented information from being passed on to others. In Communist Romania, things seemed just "to happen" in a wordless vacuum. Sometimes the incident would grow, as this one was evolving. I was amused by these incidents and accepted them without having to understand. I also felt sorry that conditions were such that Romanians found it imprudent to take anyone into their confidence.

Howard and I were on holiday, we had no rigid timetable. I wanted him to experience the Romania I was living in and this kind of event was very much part of it. It was also prudent to grasp whatever opportunity came our way. I didn't dwell on whether we were being manipulated or not. I knew that if we were, it was out of necessity. There was no malicious intent.

The gentlemen spoke to Howard and me in proficient English. 'D's turned out to be just as good. They told us that they had not met 'D' since before the War. All three friends were from Timişoara, members of that city's Hungarian Jewish community. The gentlemen had studied medicine and, after graduating, had set up practice together in Sighişoara, in the 30s.

When the War began, all the players wanted access to Romania's oilfields. King Carol II appeared to vacillate between granting access to France or Germany. Ion Antonescu, sympathetic to the fascist Iron Guards, was more decisive. With political and military help, Antonescu wrested power from King Carol and allied Romania with Hitler. Many Jews and Gypsies were deported.

Along with others, these two doctors had been transported to Germany as slave labour. Freed after the War, they made their way back to Sighişoara hoping their wives and children might also return. They waited for months, years. One traumatic day, they were presented with evidence that their families had been sent to concentration camps and exterminated. The hair of one of those gentlemen turned white

The Kilt Behind the Curtain

overnight. Both remained in Sighişoara, trying to rebuild what was left of their lives. They had succeeded, they told us impassively, as well as might be expected.

By accepting Romanian's apparently unexpected plans, I gained some unique experiences.

We said goodbye to 'D's gentlemen friends and continued north. By the time we reached Târgu Mureş it was dark. As everywhere, there was little in the way of streetlights. We found a hotel. The women entered first, registered and retired. Only then did Howard and I enter and take rooms for ourselves.

Târgu Mureş was another well-preserved, medieval city with a turbulent history. It had been ransacked by the Mongols in the 12th century, then conquered by the Turks and absorbed forcibly into the Ottoman Empire. Liberated from the Turks, it became a bone of contention between Hungary and Romania.

After driving north-east, we stopped to admire the painted monasteries of Moldoviţa and Suceviţa, then continued to Rădăuţi. The next day we drove to Siret in the far north. As had my mother and I in 1968, Howard wanted to "peer into the Soviet Union" at the border crossing some miles north. 'M' and 'D' were not interested in viewing the USSR. They'd lived under its shadow and had suffered enough. We left the women at the first chapel to have been built in Romania, the tiny stone Church of the Holy Trinity, promised to pick them up in an hour or so, and drove north. We planned to get as close to the crossing-point into the Soviet Union as we dared without causing alarm.

"Nothing's happening!" Howard was dissatisfied with the inactivity at the border. "I see armed guards, the barrier's lowered, but there's no movement. Nothing!" He was disappointed.

"Is there a fence along the entire border? Or just a line of lookout towers with armed guards?"

"I'm not sure"," I answered. "My guess is that the steel-mesh link-fence at the crossing point gives way to lookout towers and foot patrols. A fence would be too costly."

"Let's find out!" Howard was determined.

Our time had expired. We returned to the Church of the Holy Trinity for 'M' and 'D' as promised.

But Howard was not to be put off looking into Russia. We found a side-road that appeared to run parallel with the border and began following it. The road entered a wood and after a few more kilometres deteriorated to a mere track. We could see nothing but pine trees. If we could find even a low rise without trees, we would probably be able to see the town of Chernivți in the Soviet Republic of Ukraine. The previous year, close to the border as darkness set in, Pearl and I had seen a glow in the sky. The Romanian border guard had explained it was Chernivți, inside the USSR.

I told Howard the perplexing history of Chernivți. It increased his desire to see it, even from a distance, and so we drove down the track, deeper and deeper into the woods.

"Surely the land has to rise! There must be somewhere we can see over the tops of the trees!" Howard was impatient. I was listening to the rocky ground threaten the car's undercarriage. No replacement exhaust pipe to be found in Bukovina!

"This looks more promising!"

Just as Howard uttered these words, I caught sight some thirty metres ahead, a camouflage net draped over a bulky object. A large-calibre cannon protruded from the net. In a flash, I was transported back to a combined military operation the Gordon Highlanders had participated in with a British cavalry regiment. We infantry soldiers were following immediately behind a squadron of tanks. The tanks formed a line, taking up battle formation to give us covering fire. Our task was to advance on foot.

When a tank adopts its fighting position, the commander orders the tank's tracks to spin in reverse. The spinning tracks gouge into the earth and dig the body of the tank into the ground until only its cannon is above the horizon. Now the tank can fire at will while presenting a difficult target to the enemy.

Once seen, never forgotten!

Battle Formation

Immediately I captured the entire picture. Only metres away, stood a group of uniformed soldiers at ease. Guns from at least three tanks pointed north, through the camouflage netting. The position of these Romanian tanks answered the question about what Romania might do if the Soviet Union invaded from the north. They would resist!

It's never a good idea to run from armed soldiers and advanced weapons. As infantrymen, we'd been trained to stand absolutely still and take stock. I immediately stopped the car and absorbed everything.

This was the only time during these two years when I thought faster than a Romanian. In a fraction of a second, I'd taken mental stock of our perilous situation.

"One, we've stumbled into at least one troop of Romanian tanks entrenched in fighting position with their guns facing into the Soviet Union. Two, four civilians in a place we should not be even close to. Three, two British passport holders in a British car with diplomatic plates; no immediate danger. Four, two vulnerable Romanian women with Romanian identification, in immediate potential danger."

"Don't move!" I uttered the order to the women. "Don't get out of the car! Say nothing! Look straight ahead at Howard's back." To Howard, "Your passport. Now!" He passed it to me. I stepped slowly out of the car passports in hand, a smile on my face and stopped.

An astonished soldier, shaving in a mirror nailed to a tree dropped his razor in a basin, said something I couldn't hear. Two soldiers covered me with rifles. One of the other soldiers handed him his battle-dress jacket. Another soldier handed him his pistol. So, he was the officer in the group! So far, so good!

I addressed him directly as he walked towards me, part of his face white with shaving soap, his battle-dress open and his pistol down by his side. He was a professional. I extended both of my arms out from my body, one hand obviously empty, in the other, two British passports. The officer issued an order and a soldier handed him a towel. He paused to wipe soap off his face before he advanced. He was, I was pleased to see, at a loss for words. Our training sergeant had drilled into us an imperative:

"In a tricky situation, always, always seize the initiative!"

"Good day, Coronel. My friend and I are lost. We are tourists. From the British Embassy in Bucharest. How can we reach Siret?" He said nothing but took the two passports with his free hand. He tucked his pistol into his belt and opened first one passport and then the other.

"You say you are from the British Embassy. You are driving a car with diplomatic plates. Why don't you have diplomatic visas?"

"This officer had as smart a training sergeant as I had!" I thought.

"You are right, Sir! This car belongs to my close friend, the British Consul. I am the Visiting British Professor at Bucharest University. My friend and I are touring your beautiful country."

"Touring? Here?"

"I understand, Sir. I apologise. Can you please tell us the best way to leave?"

"Do not leave! Wait right there! Do not move!" By now all the soldiers were armed and grouped.

The officer walked with our passports to an official looking tent so camouflaged that I hadn't noticed it. A tank regiment is normally supported by a communications unit. It appeared that this officer was in complete charge. He reappeared a few short minutes later. He had buttoned his battledress, combed his hair. To my utter astonishment and profound relief, he handed me back our two passports.

"You will turn your vehicle round right here. You will go back the way you came. When you reach Siret, you will be met by the Miliţia. You must follow their instructions."

I nodded and thanked him warmly to show him - and to try to convince myself - that I was unafraid. We were not in the clear yet. The Miliţia were the regular city and highway police but they came under the supervision of the Securitate. I got back in the car. Howard looked shaken but still had a smile on his face. 'M' and 'D', expressionless, still stared obediently at the backs of the seats in front of them.

The officer called an order to one of his soldiers who beckoned me into making the 180 degrees turn so that I didn't approach the tanks.

Then, when I was back on the track and facing Siret, he waved me on. Very, very slowly, I drove ahead until the soldiers and the camouflage netting covering the tanks were out of sight. Then I drove faster. Nobody in the car said a word.

Fifteen minutes later as we entered Siret, two *Milițieni* were waiting for us, their right hands raised in warning. I stopped. They approached. None of us had the slightest idea what to expect. Would we be taken away for questioning? Arrested? I could feel that 'M', 'D' and Howard, were as tense as I was.

The Milițieni indicated we should wind down all four windows. Unsmiling, they peered inside.

"Get out! Open the trunk!" I did as they asked. They tapped the trunk, the wheel wells and opened the four small suitcases.

They gestured me back into the car. "Drive south for 5 kilometres. At the junction, take the road marked Dornești. Do not stop or change direction before you reach Dornești! The Milițieni there will give you further instructions."

We breathed a collective sigh of relief, drove to the fork in the road and took the branch for Dornești.

As we approached the village, two more *Milițieni* met us and repeated the same procedure.

"Continue driving on this road for kilometres until you reach Rădăuți. Do not stop for any reason. The Milițieni in Rădăuți will give you instructions."

Two Milițieni officers were standing in the middle of the road at Rădăuți. We stopped. They counted the four occupants of the car, checked the wheel-wells and the trunk and ordered us on.

"You must continue in the direction of Bucharest."

"Can we stop before we get to Bucharest?" I wanted clarification. Our vacation trip was far from over.

"So long as you continue in a southerly direction you may spend the night where you want. When you find a hotel, check in with the local Miliția."

As far as Howard and I were concerned, the trip may not have been over, but 'M' and 'D' were so shaken they asked if we could make it to

Bucharest that night. It was over 500 kilometres to Bucharest. It would be dark in a few hours. I never drove in the dark because Romanian trucks had only high-beam headlights that could not be dipped. When they passed each other in the dark the protocol was to switch headlights off altogether! This dangerous practice terrified me, so I stuck to daylight driving.

I explained this to the women. They understood. The shock they had suffered followed by the anxiety of not knowing what might have happened to them had any one of first the Army and then the Miliția asked to see their papers, was only now beginning to wear off.

We spent two fascinating and enjoyable days in the ethnic Hungarian region of Ciuc. 'M' and 'D' took on the task of hunting down the embroidered cushion-covers that I wanted. 'D', with some considerable difficulty, was able to find me four beautiful hand-sewn cushion-covers that I still treasure.

Days later, Howard and I dropped 'M' and 'D' off at Bucharest's Gara de Nord. Howard and I returned the car to the Embassy compound and parked it where the British guard told me to. He took the keys.

"How did you two gentlemen enjoy your sightseeing trip?"

"Very well," I told him, "Romania's a fascinating country."

"Did you enjoy Predeal? Brasov? Poiana Brasov and Bran Castle?"

"We certainly did!"

"And what about Sighișoara? Târgu Mureș? Bistrița?"

"Transylvania is fascinating." Howard talked enthusiastically about the mountains and forests, the peasant villages, and the ancient medieval towns.

"The painted monasteries? My wife and I want to see these before this tour's over!" Bob persisted.

"They're beautiful. Don't miss them, Bob, your wife will love them!"

"And Siret?"

"Northern Bukovina is one of the loveliest parts of this country," I told him.

"You must have really liked Siret. You were in and out twice. Twice to Dorneşti, Rădăuţi."

Bob looked at us, a smile on his lips. He was playing with us.

"Curious how I know?" He asked lightly.

"I filed my itinerary before I left." But both he and I knew I hadn't listed all of these places on my itinerary. He knew we hadn't even reached the Delta.

"Stick to your itinerary then, Ron?" A rhetorical question.

Bob beckoned us into his office. He showed us the trip itinerary I'd filed. Then he opened a bound ledger and put his finger on one line after another – place, date; place, date; place, date. Howard and I looked from the ledger to each other and back to the ledger again.

"Every time a diplomatic plate is seen by a Miliţia officer anywhere in this country," Bob continued, "the sighting, including the plate number, time and location is reported to the Secret Police. If the plate is British, Securitate immediately calls me here and gives me the plate number. I must tell them if the car's in authorised use and who's driving it."

He paused and looked at me. "I told them you were driving with some British friends and that you were following the precise route you filed with me." He winked and closed the ledger.

"OK, Jock, you and Mr. Owens can go." The NCO dismissing the rookies.

Bob was a man after my own heart. He probably wished that his stay in Romania could have been as exciting as mine obviously was!

And Howard? Howard told me that, back in London, he dined out on his adventures in Romania for a long time!

38

MORE TRAVELS WITH PEARL

Cap de Crap

A fresh spring morning. Pearl and I are driving north-west across the Wallachian Plain. We left Bucharest half-an-hour ago. We plan to spend a couple of nights in Horezu, an ancient convent I've visited several times.

First, however, I want Pearl to get a glimpse of Ploieşti and then enjoy climbing slowly into the Argeş Valley to reach the ancient town of Curtea de Argeş before turning west to the city of Râmincu Vâlcea and finally heading west on country roads to Coteşti. The countryside is in bloom.

We dutifully avoid Ploieşti, whose oil fields are off bounds to foreigners. We can appreciate the size and complexity of the oil refineries better from a distance. We remark on the massive holding tanks, refraction towers, steaming, coolers linked by miles of beautifully marshalled tubes and pipes.

In August 1943, the Americans had bombed the Ploieşti refineries because they were in the hands of Germany. In those raids, the US lost over 50 bombers and more than 500 airmen. Now, in 1969, the oil extracted and refined at Ploieşti was Romania's principal dollar earner.

We began to ascend into the spring-fresh hills flanking the Argeş Valley. They were white with plum blossom. The peasant holdings were tiny wooden houses, each in its own vegetable and flower gardens bounded by picket fences. Stout women and sturdy men tended to their gardens and orchards.

I told Pearl about the hospitality of peasants and she suggested we stop. "We will," I promised, "but later, on the road between Curtea de Argeş and Râminicu Vâlcea for lunch."

Curtea de Argeş in the southern portion of the Fagaraş Mountains had been the capital of Wallachia in the 13th century. Then, the region was divided into principalities ruled by governors or princes. We wanted to visit the magnificent byzantine church in the town and to get the 'feel' for what the original capital of the Principality of Wallachia had been like, but we found the road blocked. A public holiday! We parked the car and walked. Men, women, and children in beautiful regional costumes were enjoying the festivities. We joined them.

"Where's the best place to have lunch?" A gentleman directed us to a narrow street on the edge of town. There were several restaurants on one side of the street and crates of chickens, geese, hens, and lambs on the other. People crowded round the crates inspecting the live options available for dinner!

Some peasant entrepreneurs were selling live geese and chickens. We could see others scoop live carp from tanks and hand them to eager purchasers. There were un-weaned lambs. For the customer unwilling to clean or pluck, a group of women would prepare the fish or fowl for a few extra Lei.

My cousin had been a shepherd on hill-farms and deer-forests in Scotland and both Pearl and I knew lambs and sheep well. However, we were unprepared for how these lambs were being sold. A customer would point to a lamb with all 4 legs tied together at the hooves. The vendor would slit the lamb's throat, bleed it into the gutter, then skin it and hand the tiny carcass to the eager customer.

Pearl and I, like the Romanians around us, accepted what we saw. Nevertheless, there was something that struck us as pathetic about seeing such tiny lambs killed. They barely had any flesh on them.

We found a table in the restaurant identified as "best". The waiter brought us menus. Of course, they were in Romanian, so I began reading aloud to Pearl and translating the names of the dishes.

"Cap de crap!"

"What?" Pearl roared with laughter. "Repeat that?"

"Cap de crap!"

When Pearl and I stopped laughing, she asked, "What on earth is "cap de crap"?

"I don't know but it's not what you think it is!"

Pearl decided on the lamb to see how such tiny animals were cooked and presented and to taste something she'd never imagined eating before!

"Why don't you order the "cap de crap" Ronald?" Pearl went off into gales of laughter again. Always one to try something new that's what I asked for.

"O porție de cap de crap și o porție de miel, vă rog!" I managed to keep a straight face.

Pearl's lamb arrived first. It looked and smelled delicious, cooked with fresh vegetables and garlic.

Then the waiter motored over to me balancing a deep soup-plate expertly above his head in one hand. He swept it onto the table without spilling a single drop or soiling the rim of the plate.

Pearl and I peered at it. Out of a red soup that offered an appetizing aroma, rose a large fish head, eyes wide, mouth gaping to show large pharyngeal teeth. Involuntarily, we recoiled. Neither of us had been prepared for this nor knew that any fish might possess such human-like teeth. We looked from the carp's grinning head to each other and then laughed uproariously attracting a lot of attention from amused diners. For them, carp-head soup was a seasonal delicacy!

Pearl and I shared each other's plates and enjoyed both. The amusement the cap de crap afforded both us and the other diners, added extra flavour to the meal.

The Convent at Horezu

The approach to the convent at Horezu always put me in mind of a Bruegel painting. A sweeping countryside, one or two large buildings that dwarfed mere people. High, ochre walls enclose an enormous courtyard. From outside the domes of a church and a belltower are visible above the walls. They need repairing in places and there on a wooden scaffold is a small group of workmen plastering and painting. The entrance is an arch with a belltower atop and into the arch are set ancient wooden gates.

Before we even stop, two nuns, foreshortened by the grand archway, begin to swing open the massive gates. We can now see the nuns' apartments soaring two stories high and stretching the entire length of one wall. The residence is copper-green with age and protected by two matching towers. In the middle of the courtyard stands the 300-year old chapel.

We park just in the courtyard. The doors are swung-to and barred behind us. Smiling and offering us gentle words of welcome the nuns escort us to meet Mother Superior Tomaida.

I know from previous visits that there are no more than a dozen nuns in the entire monastery. All are elderly. When the Communists took over in 1947 the monastery was allowed to remain open, but no novices could be admitted, so the residents steadily aged and their numbers declined.

Mother Superior Tomaida welcomes us and expresses pleasure at meeting my mother. We will be escorted to our rooms on the second floor but first, the evening meal in the communal dining room. The sound of wood-on-wood tells to take our places, sit, and say grace.

Orthodox churches in Romania use not a bell as a summons but a *"semandron"*. In appearance, a semandron is like a light, double-ended, wooden paddle. Holding this instrument in one hand, the nun beats on the flanges of the paddle with a wooden drumstick held in the other hand to make a series of staccato beats. A priest or nun skilled in the use of the semandron – onomatopoeically called a *toacă*, pronounced *twaka*, in Romanian – speeds up the rhythm to give an increased sense of urgency as the limit approaches for whatever ritual is being announced.

Pearl was charmed by the dining room. It was as old as the chapel and its walls adorned with religious frescoes any church would have been proud of.

We and the nuns all sat down together. A simple meal of "çiorba de urziçi", nettle soup. Pearl and I were offered bread, but the nuns took only the soup. Out of solidarity, we too decided to forego the bread. Mother Superior Tomaida told us that the nettles had been collected that afternoon by two of the sisters. Today was a religious fast day, the nettles being symbolic of abstinence.

During the War at my grandmother's house in Coupar Angus we had regularly picked and cooked nettles and so we found the soup delicious to the evident satisfaction of Mother Superior Tomaida and her frugal nuns. They received few visitors, had been looking forward to our arrival but were concerned that we were arriving on a fast day. Our empty plates relieved them of any concern for our nutritional well-being.

"Later this evening there will be a full mass celebrated in the chapel by a visiting priest. Would you care to attend?" Mother Superior Tomaida inquired. Both Pearl and I eagerly accepted. Gratified, we were to listen for the rattle of the *toacă*.

When the *toacă* sounded several hours later, we were dressed and ready. We made our way down the ornate stone stairs of the residential block, crossed the dark in darkness and entered the chapel. Although the chapel was large, it was lit by only a very few candles and its walls and ceilings were entirely covered in frescoes depicting lessons, giving the impression that we were in an intimate space.

A curved wall faced the iconostasis, a magnificent, carved screen decorated with icons that separated the rounded nave, where we and the nuns sat, from the sanctuary with the altar and the Eucharist. In a Romanian Orthodox church, the sanctuary is accessible only to the priest.

We sat in silence. The nuns' calm, aged faces appeared centuries-old in the semidarkness. Eyes and haloes of the Virgin and Saints

The Kilt Behind the Curtain

glinted from the shadows that licked us as the candle flames flickered. The ancient, magnificently bearded priest lit three enormous candles before the iconostasis and the icons sparkled in bright silver and warm gold.

We watched an ancient haunting rite conducted in Old Church Slavonic, a beautiful thousand-year-old language of God's love. The splendidly robed priest was more active and vocal than any Church of Scotland minister. He energetically swung the gleaming censer to produce gusting clouds of fragrant incense accompanied by bell-chimes. He intoned the liturgy in a deep, clear voice. The nuns chanted simple responses as the ritual progressed.

We left the mystery and the colour, the tradition and the ritual of the chapel for a beautiful starry Carpathian night. In silence, we breathed in the fresh, sweet mountain air that bore the aroma of herbs and blossom and lowing cattle. Deeply moved, we climbed the wide stone staircase to our austere, spotless rooms in the residence.

Harvesting willow-whips for basket-weaving. In the country, people were friendly.

39

DANUBE DELTA

Tulçea and Further Afield

The Danube Delta is different from other parts of Romania. Driving across the Dobrogea Plain to the horizon gives the Delta region a feeling of isolation. Pearl and I were on our way north from Constanța to Tulcea in a borrowed Ford Anglia.

Two days earlier, we'd driven from Bucharest to Cālāraș, taken a ferry across the Danube and continued on to Constanța, Romania's principal port on the Black sea. We'd stayed in a private house whose address we'd been given by the Romanian tourist office. We knew from experience that hotels in Romania were impersonal and we hoped that state-approved bed-and-breakfast might give us more opportunity for contact with Romanians.

Our hostess, Doamna Lupei, a woman of Pearl's age, turned out to be a pure delight. Her husband was a mariner on the Danube and was gone for days at time. She too, hoped that offering bed-and-breakfast to Romanians who came to enjoy the beach might help her pass the time pleasantly. She had never anticipated playing hostess to two Scots, one of whom spoke passable Romanian!

Pearl and Doamna Lupei immediately became friends although

neither spoke the other's language. They communicated using their hands, facial expressions, and much hilarious laughter.

We asked Doamna Lupei where we should go for an evening meal. To our delight, she offered to make us mămăliga cu brânză – polenta topped with butter and crumbly cheese. She was showing us how to prepare the polenta when a thought struck her.

"I'm not sure that the Tourist Office would approve. I'm probably doing a local restaurant out of business". A Romanian wouldn't dare run afoul of the authorities even by accident.

"Why don't I go out and buy a bottle of Murfatlar wine and some pastries and bring them back. Then we will have done our duty by the local businesses and I'll have the receipts to prove it."

This satisfied her. I left them chattering away as if they'd known each other forever.

The polenta, butter and cheese couldn't have tasted better. The Murfatlar complimented the dish. I had bought pastries and fruit for dessert as well as bread and sliced ham for the following day's breakfast just in case an informer was watching my movements.

After dinner, we sat with Doamna Lupei, with the windows open and the sounds of the busy port as background music. She showed us mementoes of journeys she had made up and down the Danube with her husband before they'd had children.

Our plan was to drive to Două Mai the next day and spend the night in the little ethnic Russian Lipoveni compound where I occasionally spent a weekend. Seeing Pearl and Doamna Lupei get on so well I suggested we visit Doi Mai for the day and return to Constanța for a second night.

We drove down the Black Sea coast and visited villages along the way, Eforie, Costineşti, and Mangalia, until we arrived at the tiniest village of all, Două Mai. The family I usually lodged with invited us to stay the night, but we excused ourselves claiming lack of time. The women of the family were delighted so show Pearl their tiny spotless cottages and their embroidered bedspreads and cushion covers. There only ever seemed to be women and children around that compound. The men might all have been away fishing. No explanation was offered and in Romania it was always better not to be too inquisitive.

They insisted we stay for lunch and baked fish for us. They cooked their fish whole, the way we did in Scotland, so that the flesh absorbed the flavour of the bones and head. We each ate an entire fish leaving only the bare bones on our plates, picked perfectly clean of flesh. We all admired our tidy plates and then smiled at one another, recognising that we shared the culture common to all fishing communities.

On the way back, we leisurely revisited the villages and the beaches and took time to admire the statue of Ovid that stood in one of Constanța's squares. The poet Ovid had been banished from Rome to Constanța. Then, the port was called Tomis and located in that part of the Roman Empire called Dacia. Ovid died in Constanța without ever revisiting Rome.

Details like this constantly reminded us how ancient a part of the civilised world Romania was and how fortunate we were to be able to explore it. Ovid's statue was placed in the square only at the end of the 19th century, but it was a reminder that Romanians loved to proclaim their Roman heritage.

While we were examining Ovid's statue, Pearl grabbed my arm and pointed in excitement to two Turks. Each wore his conical red fez with black tassels. One of them was wearing shoes with upturned points at the toes. The Muslim Turks had colonised Dobrogea from the mid-16th till the mid-19th century and brought it under the Tomes Ottoman Empire ruled from Constantinople.

Doamna Lupei welcomed us back into her home and again invited us into the kitchen as she prepared dinner. She stewed onions and carrots, added cooked rice and a little ground meat to make meat balls. Then she rolled the meat balls in pickled vine leaves to make small, wrapped packets called *'sarmale'*. She steamed and then served them smothered in thick plain yogurt. We enjoyed these with another bottle of local white wine and ended the meal with sweet pastries.

After dinner, Pearl explained to Doamna Lupei, with many a bizarre gesture, that we had seen two Turks in traditional dress. Doamna Lupei mentioned that there were two mosques in Constanța,

The Kilt Behind the Curtain

an older one that was closed and a newer one that was still in use. She offered to take us to the closer one. It might have been an Orthodox church were it not for the distinctive minaret. The Romanian Government had built it before World War One as a goodwill gesture to the Turkish Muslim minority in the city.

The following morning, we had to drag ourselves away from Doamna Lupei. This would be our last opportunity to visit the Danube Delta, an experience we didn't want to miss. We drove towards Tulcea, fifty miles inland from the Black Sea but still one of Romania's important ports on the blue Danube.

It was early when we reached the port. The road led us directly to the quays on the riverbank, quays crammed with storage sheds and cranes for loading cargo. It gave Pearl and me a thrill to look out across that wide river and realise that its waters began almost 3,000 kilometres to the west and flowed through Germany, Austria, Czechoslovakia, Hungary, Romania, Bulgaria, and finally shared its enormous fan-like delta with the Ukraine now in the USSR. A major European river that started in the Black Forest and ended in the Black Sea not far south of Odessa. It was navigable for most of its length.

We were planning to reach Sulina, the Danube's final port at the furthest point east, where the Delta met the Black Sea.

My friend and colleague Dino Sandulescu had told me an intriguing tale about Sulina. His father, Nicolae, had been born in there in 1891 and had later moved to Bucharest where he married and became a banker.

Dino told me how he'd been taken to Sulina in 1937 when he was barely 5 years old to meet his paternal grandparents. His grandfather held a senior post with the Danube River Commission whose headquarters were in Sulina, then a bustling town of 6,000. He spoke Greek, French, English and Romanian. However, Dino's grandmother spoke only Greek. Dino distinctly remembers having been coached in that language for some weeks before making the trip so that he would be able to greet and converse with his grandmother in her mother tongue.

That story that conjured up elements of romance: a multicultural, multilingual family; an isolated corner of Europe but of such commercial importance that it housed the headquarters of the Danube River Commission; a child's swift acquisition of modern Greek in order to show respect for his grandmother's Greek heritage; an arduous journey south by train to meet the Danube and then north again by motor-launch for three days in a sweeping curve all the way to the immense reed-jammed waterways at the mouth of Europe's longest navigable river, to Sulina itself. Sulina, a mere 300 kilometres from Crimea where the ignominious Charge of the Light Brigade had been led against the Russians at the Battle of Balaclava in 1854.

There wasn't a British boy or girl in my school days who couldn't recite Tennyson's The Charge of the Light Brigade.

"Forward, the Light Brigade!"
...
Theirs not to reason why,
Theirs but to do and die:
Into the valley of Death
Rode the six hundred.

We heard that spellbinding poem recited to us at the Morgan Academy when we were seven years old.

The opportunity to observe wild fowl was an additional motive to visit the Delta. However, try as we might, we could not find the road to Sulina and so I asked a wharf-side crane-operator.

"You want to drive to Sulina, Comrade?"

"That's right, but I can't find the road."

"Don't you have a map, Comrade?"

"I do!" I spread it out and traced the thin blue line, that I imagined was the secondary road that joined Tulcea with Sulina.

"That's not a road, Comrade!" He was grinning. "That's a channel. The Sulina branch of the Danube. It's one of the navigable routes that allow cargo boats to reach the Black Sea."

"The only way to get to Sulina from Tulcea is by boat?" I was taken aback.

"Only by boat, Comrade!" He smiled at my surprise that the only viable means of transport in the entire 4000 square kilometres of the Danube Delta region was by boat!

We were disappointed that we couldn't drive through the Delta to the coast as we'd planned but tried the next best thing. We followed a narrow road that ran parallel to the most southerly of the three navigable channels – the St. George Branch – and decided to see how far we could get.

We reached a small village called Malcoci, just a few kilometres south-east of Tulcea. Malcoci was a collection of neat wooden houses surrounded by picket fences. It was so attractive that we stopped on the outskirts, parked, and walked back to look at the beautiful flowers growing in one of the gardens. Then the cottage itself caught our attention. Its eves sported decorative woodwork supports and out of each had been cut the shape of a fish through which the sky shone clear and blue. As we were admiring the careful carpentry work, the owner – an old woman – came out of the open front door and greeted us. I explained who we were – Westerners driving a car with diplomatic plates, just so she could decide whether she really wanted to talk to us or not. Apparently, she did.

She opened the wooden gate, invited us into the garden, guiding us around naming the flowers she was proud of including tall hollyhock that were so dark as to be almost black. Her husband was tending their vegetable garden. She introduced us to him. He was from Sfântu Gheorghe, a tiny port at the mouth of the St. George branch of the channel. He had spent his working life on barges transporting grain from the hinterland for exportation to far-away places.

Together, they invited us to see the inside of their home. It was similar in size to the farm-workers' cottages in Scotland – two rooms – but how different inside! It was light, airy and dry. In one corner stood a ceramic stove that had a broad extension with a mattress and bedding on top to provide a warm place to sleep. No slipping between damp sheets in a cold room as we tended to do in Scotland.

Ceramic stoves were everywhere in Romania. They provided a practical and far more efficient system of heating than open fireplaces.

Interior flues drew all the heat from the firebox through a maze of internal tunnels. The result was that the surface of the stove never became too hot to touch. Every ounce of heat was absorbed into the interior firebricks and percolated comfortably though hand-painted tiles.

From the tiles of the stove to the bedcovers to the table and chairs and even the walls, there was colour. Bedcovers and tablecloths were hand-embroidered; the simple table and chairs bore hand-painted patterns of flowers and fish.

Pearl and I remembered seeing one of these ceramic stoves being built from scratch in Horezu Monastery while the craftsman explained the logic of the technology to us. The purpose behind the stove was to extract all of the heat through a labyrinth of flues built from refractory brick so that the heat would be enjoyed inside the room.

Equally intriguing was how the craftsman made and painted the tiles. He shaped them by hand, let them dry a bit and then used a cow's horn as an outsize fountainpen to sketch patterns in rather drab colours. He then baked the tiles in an oven and when they were removed, they turned out to bear intricate designs in many beautiful colours on shiny surfaces. The pigments had changed colour in the heat. The artisan used these tiles to clad the firebricks and so give dignity and beauty to the finished exterior.

On one wall of this cottage hung Orthodox Icons of various sizes. The largest was made from a single slab of wood that had warped and curved into an arc as the slab had dried. It was both dignified and bright. The Saint's head was encased in silver with thick, embossed lines radiating from it as if from the sun. There were several tiny wooden icons and one delicately encased in silver.

In the countryside far from bureaucracy, peasants were proud to show us their orchards and gardens, their geese and domestic animals, even the interior of their houses. So different from Bucharest.

Given what the old man told us about the condition of the road ahead we decided to backtrack to Tulcea and drive inland to spend the night in Brăila, a larger and more important port than Tulcea.

All I Really Want...

On our way to Brăila, I reversed the car off the road into a field entrance to take a photograph. A farmer on a tractor was cultivating half-a-kilometre away. The sun was setting. Pearl and I got out to admire the orange ball sinking slowly out of sight. I wondered if I dare photograph directly into the setting sun. It was taking me a long time to find the right exposure. Suddenly Pearl warned me that the farmer and his tractor were now making a beeline towards us.

Fast thinking was called for. There was no reason why he should object to us being parked in the field. A Securitate officer might wonder why we were taking photographs, but a farmer should not interfere. Pearl returned to the passenger seat and closed the door. As I closed the driver's door, the farmer came running over waving his arms.

"Please don't leave!"

"Do you want me to take a photo of you and your tractor?" I asked.

He looked puzzled. "All I want is a bicycle with a dynamo!"

I thought I hadn't heard correctly and asked him to repeat what he'd said.

"All I want is a bicycle with a dynamo!"

I stared at him. It was as if I'd started watching a film in the middle of a wacky conversation.

"A bicycle?"

"With a dynamo!" He insisted.

Thinking I'd misunderstood I asked, *"You have a bicycle with a dynamo?"* My mind raced and I wondered if he was telling me that his bicycle had broken down or his dynamo wasn't working and, because it was getting dark, he wanted us to offer him a lift back to the collective farm.

Annoyed, he shook his head. "You're from the West!" He gave a meaningful glance at the car so that I could not deny the fact.

Slowly and distinctly he repeated, "Comrade, what I want is a bicycle with a dynamo. That's all!"

It dawned on me that he was serious. Romanians tended to believe the assertion that "Westerners have everything; probably two of everything!" Even sophisticated acquaintances in Bucharest would offer me an unwanted gift in order to put themselves in a position to make a request of me. I'd been dumped with a huge rubber plant, a 10-gallon demijohn of undrinkable wine and a box of Russian books. Once I'd "accepted" the gift I was expected to return their "generosity" with some favour they were in desperate needed of. Failure to comply would bring accusations of my being disobliging.

This farmer was going one step further. He was counting on getting his bicycle, with lights, free and he wanted it now! He seemed to believe that Westerners ran some sort of "cargo cult" that allowed them to deliver whatever was asked for.

I shook my head. "I'm sorry, I don't have one."

Had my mastery of Romanian been better I would have liked to have explained why I was simply not in the position to satisfy his heart's desire. But I didn't and even if I had, I doubt if he'd have believed me. After all, Westerners were affluent! Every communist in Romania knew that!

He gestured towards the boot of the car. Only a child's bike would have fitted there. I opened it. He shifted our suitcases. Disappointed, he gestured towards the car's interior. I opened the doors and he peered at the empty space for a long time.

"I'm really sorry, comrade." I shook his unwilling hand. We bumped back onto the road and drove away leaving that dejected figure standing there in his black căciulă looking, and no doubt feeling deceived that the first Westerner that he'd had the good fortune to meet, had been unable to accommodate his heart's desire. I've often wondered if that experience embittered him against the West for life!

40
SIEBENBÜRGEN

Kronstadt

Harald Mesch and I became friends and we saw each other regularly in Bucharest throughout the fall, winter and spring of 1968-69. Harald would speak relatively openly to me in the Faculty Common Room. He also visited me in my apartment.

As I did with all Romanians, I let him make his own decisions on these matters. I'd come to the conclusion that only Romanians themselves could fully understand their political environment, assess the risks they ran and determine how best to deal with them. It was wise for me not to initiate friendships but if a Romanian showed that he or she wanted to associate with me, I had to assume that they knew exactly what they were doing and were willing to take responsibility for their actions.

Since the Communist Party had grabbed power in 1947, Romanians had developed a keen sense of what could or might not get them into trouble with the secret institutions of the State. They taught their children these essential survival skills even before they taught them to read and write. Being constantly on the watch, constantly suspicious of neighbours and friends, must have played havoc with personal relationships. This saddened me but was beyond my control. All I could do was to demonstrate my openness. The other party was then free to take tiny steps towards me until they reached the depth of relationship they felt comfortable with. Some relationships might only reach to a smile and the exchange of a few words in the Faculty Common Room or they might blossom into a deeper friendship as they had done with Dino and Harald.

I suspected that some of my university colleagues would like to have befriended me but for reasons of their own, reasons that were impossible for me to imagine, they kept their distance. I respected that.

One excellent example was Professor Mureşanu. He was in his mid-30s and gave the impression of possessing great energy and enthusiasm. If he happened to be in the common room when I entered, he would lurch forward as if to greet me but abruptly draw back as if he'd suddenly remembered some prohibition. It was embarrassing. Nevertheless, my only option was to respect the retreat, not to encourage the advance.

One day after I'd finished delivering my classes, I crossed the boulevard to catch a bus and joined the loose group already waiting. Mureşanu was among them. Our eyes met. His face lit up. He lurched forward grasped my hand and pumped it as if we were long-lost friends. "Professor Mackay! Such a pleasure!" Suddenly, he disengaged and pulled back, enormously embarrassed as if aware of his public recklessness. The solution was for me to end his embarrassment and so, with a "See you later!" I left and started to walk.

Moments later, his bus passed me. Professor Mureşanu was next to the window. He looked miserable. Although our eyes met, neither of us exchanged any sign of recognition. Years later, I heard that he had defected and gone to the United States to become a cowboy or a rancher. I truly hope that Mureşanu never again has been forced to reign in his enthusiasm for life.

Harald Mesch had relatives in West Germany with whom he was in regular contact and from whom he received gifts from time to time. One of these was an eye-catching winter coat that had the texture and appearance of a European brown bear. It drew attention to him, but he seemed not to notice.

Bucharest was not his home; he was from a region he called *Siebenbürgen*, the Seven Cities. It had been given the name Siebenbürgen, he told me, after the seven towns that Saxon settlers had built in the early Middle Ages – Bistritz (Bistriţa), Kronstadt (Braşov), Klausenburg (Cluj), Mediasch (Mediaş), Muehlbach (Sebeş), Hermannstadt (Sibiu) and Schäßburg (Sighişoara). Only when I naively commented that it must be close to Transylvania did he tell me that Siebenbürgen was the original German name for what was now known in Romania as Transylvania. As a Scot I could identify with his proud feeling of belonging.

When I told Harald that I loved the Carpathians and showed I knew some of the mountainous areas fairly well, he invited me to spend a weekend with him in Braşov – which he always referred to as Kronstadt. He told me that he'd been born in Hermannstadt (Sibiu) but brought up in Kronstadt by his maternal grandmother, his *Oma*, after his mother had died when he was an infant during the Second World War. We spent most of the weekend with his grandmother although I slept in separate accommodation rented through the State tourist office.

It was clear that they had a very great affection and respect for each other and that his *'Oma'* was proud of the teaching position he held in Bucharest University. They spoke to each other in *Sächsisch*, an early German dialect that was current in the region.

Oma must have been warned well in advance that a Scottish friend was coming to visit, because every meal was designed to introduce me to local Saxon cuisine mostly made from scratch at home. She made sauerkraut and we ate it cooked in beer and eaten with pieces of smoked ham; we drank her clear, pale, straw-coloured, home-made elderflower wine and ate her *lebkuchen*. For the very first time I ate *schpeck*. It had been fried and looked, on the plate put before me, for all the world like a very thick and very solid slice of pig fat with the orange skin still clinging obstinately to it though fortunately well-shaved of hairy bristle.

"Saxon bacon!" Harald announced proudly; Oma smiled acknowledging that this too, was local.

'Bacon?' On close inspection, I could make out a thin streak of red bacon through the middle of the fat. Judging by the way *Oma* and Harald enjoyed theirs, *schpeck* was a delicacy. To show my appreciation, I ate mine with gusto. In those days we'd never heard of cholesterol and I possessed teeth capable of reducing anything to a point at which it could be swallowed without choking me. I told them how delicious it was. *Oma* asked if I wanted to see where it came from. Hoping for an excursion to a farm, I said *Yes!* She led me to the window and pointed to the second-floor rear balcony of a neighbouring block. I followed her finger, puzzled. She laughed and said something to Harald.

"She says you have to look very closely!"

I looked very closely and sure enough, in a large steel cage an enormous pig was lying right there on the balcony.

"These neighbours are *Oma's* friends. They produce a fat pig every six months. She helps them slaughter. Certain cuts are pickled, others dried, the hams are smoked. *Oma* shares in all the benefits.

I questioned the practice of raising farm animals in an apartment block. *Oma* pointed out that many Saxons had moved to town from small farms. With them, they brought their traditions of self-sufficiency. From her window we could see cages with hens, geese and rabbits on neighbouring balconies.

Harald invited me to Hermannstadt to spend the last night of 1968. Liesl, his wife, was ill, at home with her parents, in the care of her father, a doctor. I was not expected to pursue the matter. I spent New Year's Eve in the company of Harald's father and a group of his father's friends. That New Year's Eve I learned a great deal about Harald and about the Saxon population of Siebenbürgen.

From time immemorial, the politics of what today we call Romania have been complicated. For centuries the region was fought over by

Mongols, Tartars, Romans, Magyars, Slavs and Turks. It was ruled by war lords and at times by Emperors. In the 19th and 20th centuries, the survival of the more fragile states depended on agreements with superior powers like Germany and Russia, Britain, and France.

In 1938, the Allied powers accommodated Hitler's annexation of the part of Czechoslovakia known as the Sudetenland, until 1918 part of the Austro-Hungarian Empire. The Soviet Union annexed parts of northern Romania in 1940. In 1941, Germany invaded the USSR and Romania. Romania wanted its northern territories back from the USSR; Germany would support Romania because it needed Romanian oil. Antonescu, the Romanian Prime Minister who had removed King Carol II in 1940, signed an alliance with Germany and attempted to recover its northern territories from the Soviets.

Romania failed to do so, and it was forced to sign a peace treaty with the USSR in 1944. Many ethnic Germans fled Romania to avoid being imprisoned by the Soviets as enemy collaborators with Germany. Their numbers included Harald's father and his friends. When the war ended in 1945, many Saxons who had served Germany returned to Transylvania to reunite with their families. Harald's father found that his wife had died. He and thousands of Saxons were rounded up and transported north into the USSR where they were used as slave-labour to rebuild destroyed cities or to construct new ones.

When Harald's father was eventually released in 1951 and returned to Hermannstadt, he tried to resume a normal life, but he had lost his wife and his son was being brought up by his grandmother in another city, Kronstadt. Kronstadt, by then known as Braşov, had had its name obligatorily changed yet again to "Stalin City". I could understand why Saxons preferred to use the Saxon name, Kronstadt.

Over tiny glasses of *ţuica*, I learned all this and more from Harald's father and his friends. The *ţuică* helped lubricate the communication which was conducted in my limited Romanian. If I failed to understand, they helped out in their far more limited English. I was surprised that they spoke any English at all. They explained.

"After the War ended, we became prisoners of the Americans for four months. In four months, we learned a lot of English. We were in Soviet camps six years, but we didn't learn a single word of Russian!"

Proud of their linguistic achievements, they laughed uproariously and poured another round of *țuică*.

Now I was better able to understand Harald. I could appreciate what hurts, disappointments, comforts, and joys had gone in to forming this careful, private, sensitive man.

However, for Harald, there was more – much more – than the effects of a childhood disoriented by the turbulence of international war. Liesl, his wife, was dying of a rare form of leukaemia. She and Harald lived with her parents in their home in Hermannstadt, where they all cared for her. Harald travelled from Hermannstadt to Bucharest by train to give classes and fulfil his professorial duties. Every weekend he returned to Hermannstadt. If Liesl was altogether debilitated by her treatments, Harald would spend the day at his *Oma's* house, working very hard so that he could spend more free time with Liesl when her suffering lessened.

Harald's relatives in West Germany had, recently, sent him a Volkswagen to make it easier for him to cope with all his trying responsibilities. Agonizing as his life was, he never displayed his pain openly. Nobody could have guessed the crushing nature of the demands made on him.

Harald had shown me where he lived when he was in Bucharest. It was in a large room in an apartment. The room was divided into tiny sections by sheets hung as curtains. In each was a single bed. Beside Harald's bed was an attaché-case that served as a table. The case and his books went back and forth with him every week. Two clean shirts hung from hangers on the wall.

I was the only one of Harald's colleagues who knew his personal circumstances. Nobody else had any idea of how wearisome his life was and the super-human efforts he had to make every single day.

About Harald

Now, in the spring of 1969, Harald was inviting Pearl and me to Hermannstadt to meet Liesl and her parents. Before we left, he quietly told us that he and Liesl had agreed that we should know what to expect. Liesl's condition was terminal. She was as well as she would ever be. Her father had already made plans to take her to West Germany. There, she would live just a little longer than if she remained in Hermannstadt. Liesl did not want us to treat her as an invalid.

Pearl and I had few words, but we understood. In Scotland, we faced mortality in a similar way.

Harald, Pearl and I left Bucharest for Hermannstadt in Harald's newly arrived VW Beetle. I was driving since he hadn't yet obtained his license. We arrived early in the afternoon, exactly the hour that Liesl and her parents were expecting us. Their comfortable apartment was in an older building that offered a panoramic view of the stark Făgăraş Mountains.

We were welcomed by Harald's mother- and father-in-law, an educated, professional couple in their mid-50s. Although they greeted us with smiles the weight of their daughter's illness was evident in their faces. They embraced Harald first; there was obviously great affection there. Pearl, their own age, was given a special welcome.

Liesl rose to greet us in the comfortable sitting-room-cum-study, a tall young woman, my age, slim and elegant. She had been resting on the divan contemplating the mountains. She and Harald embraced. Then she turned her smile on Pearl and me. Her graciousness couldn't hide the fatigue in her eyes. She had been alternately reading, resting and painting in water colours that afternoon. She showed us her work.

"Mogoşoaia!" Pearl immediately recognised the sense of stillness of the palace garden we had visited more than once. "That's one of the smaller lakes. You've captured the atmosphere beautifully!"

Pearl's compliment to Liesl's talent was the perfect start to a necessarily subdued afternoon. Liesl showed us her portfolio of water colours. Pearl told amusing anecdotes about our Romanian journeys and kept the mood light-hearted.

When we took our leave to go to the bed-and-breakfast that I had arranged through the State tourism office, Liesl presented Pearl with the water-colour of the palace at Mogoşoaia. Forever after, wherever Pearl lived, the water colour went with her and was given pride of place on her wall. My mother had been deeply moved by Liesl's circumstances. She admired the dignified way the family handled them. She felt deeply for the gentle Liesl, her parents and for the sensitive, uncomplaining Harald.

Shepherds Watch Over Their Flocks

We drove back from Hermannstadt down the banks of the Olt River with the forested Carpathians on either side of the valley. We were in no hurry, each engaged in our own thoughts and enjoying the lush, early summer countryside as it gave way from mountain to plain. Just south of Râmnicu Vâlcea, a vast flock of sheep spread out ahead of us, blocking our progress. We were used to sheep on Scottish roads. Drivers must be alert. Sheep have priority over vehicles. However, a flock of this size, Pearl had never seen before. There must have been six thousand sheep.

In Romania when spring came and fresh grass began to flush, the shepherds traditionally led their flocks from low winter pastures to summer pastures on higher land.

Romanian shepherds and their dogs handled flocks in an entirely different way from their peers in Scotland. In Scotland, the sheepdogs did most of the work. The writer James Hogg wrote:

"Without the shepherd's dog, the whole of the open mountainous land in Scotland would not be worth a sixpence."

In Romania shepherds walked ahead of their flocks, leading them. Their surprisingly large, patient, dogs walked each with its own master. At night, however, these turned into ferocious guards threatening any animal that might prey on their sheep. They could hold their own against wolves, even bears.

I'd long wanted Pearl to see a Romanian shepherd face to face. They always looked to me as if they had just walked out of the pages of a beautifully illustrated Bible, or a medieval history book. Now, right here, she had the opportunity. We'd stopped to let the thousands of sheep mill quietly by us.

Harald and I greeted the closest shepherd and engaged him in conversation so that Pearl would have time to study his appearance in detail. He wore a tall, conical căciulă of white sheepskin. From his shoulders almost to his feet hung a cloak made of unshorn sheepskins sown together. He wore it slung over his shoulders, open in front. It was his protection from the cold night. He wore a shirt that had once been white, a black waistcoat, and a pair of home-spun trousers. His shoes were also home-made. Each was a curved piece of thick leather with a high curved point in front. They were fastened with leather thongs tied crisscross fashion over his ankles and part way up his trouser legs. He carried a walking stick and a worn leather bag.

The reek coming off his untreated sheepskin cloak was overpowering. It was probably a blessing in disguise, because his personal condition beneath the cloak, suggested that he hadn't washed in weeks. He would have a bouquet all of his own! With a week's growth on his pleasant, weather-beaten face, he grinned at us to show the odd yellow tooth. He was probably in his early 40s.

While his fellow-shepherds continued to lead the enormous tide of sheep onwards to pastures new, he was more than delighted to answer our questions about his work, his sheep, the distance that they had come – *"Several days'"* – the distance that they had to go – *"Several more days'"* – and to show us his cloak, his bag and its contents, his stick and his sandals. Pearl kept well up-wind of him. Eventually, he waved us a cheery *"La revedere!"* "Until we next meet!" and waded through his flock to take up the lead position, with his fellow-shepherds and their dogs.

Loss

Liesl died in West Germany a few months later. Harald was granted both passport and exit visa on compassionate grounds. Harald did not return to Romania. The Saxons of Romania are considered as "Auslandsdeutsche" – German nationals living abroad. The German Government grants them full rights of German citizens. Harald was immediately hired as research assistant to a respected German professor of American Literature. He successfully defended his doctoral dissertation, and, in his turn became a distinguished scholar and professor.

Final Recital

Harald invited me to an organ recital in one of the Protestant churches on the main boulevard. The invitation came with strict instructions. I had to arrive alone and enter by a side-door when nobody else was entering, then climb the spiral staircase to the gallery. A celebrated Saxon organist was to give a recital. When I took my place in the gallery, the church was already almost full. Everybody sat in silence.

Then a single light illuminated part of the sanctuary and a gentleman announced that the organist was offering this recital on the eve of a concert tour he had been invited to in West Germany. The organist bowed and sat at the organ. He played for about 40 minutes to a silent crowd. The music soared into the darkness.

He finished, stood up, turned, and faced those present. The audience was silent. After a long, low bow, he disappeared into the silent darkness. We left by the same side door, individually or in pairs. It took a long time. As I waited my turn, I could see many, mainly older couples, were wiping their eyes.

As I watched and reflected on the spectacle, the truth hit me. I had witnessed that young musician's last recital in Romania. Later Harald told me that his defection had been reported on Radio Free Europe.

It seemed to me that the minorities in Romania had greater freedom to defect than ethnic Romanians. The contact with relatives in the West that ethnic Hungarians, Germans, Jews, and other minorities had seemed to be tolerated by the Securitate. On the other hand, ethnic Romanians who received a letter from abroad might be called in for questioning, their intentions, and their allegiance suspect. One more puzzle to be stored away in my Western brain, unanswerable and unanswered. The most credible hypothesis I heard was that Ceaușescu's regime was taking payments from Germany to release ethnic Saxons to Germany, and it traded Jewish emigrants to Israel for hard currency and weapons.

41

IVAN DENEŞ

Getting to Know Ivan

"Ivan Deneş, a close friend of 'D's, would like to get to know you. Ivan was released from prison a few years ago. He's an ethnic Hungarian, a Jew and a member of the Writers' Union. Would you be willing to meet him?" Could I refuse, given 'M's engaging sketch? 'D' had, it seemed, interesting friends. And so, a meeting was arranged.

'M' duly told me that I should sit at a table in the Herăstrău Park restaurant. Outside tables had been removed for the winter, but the heated interior was open. She told me the day and time and that I should remain seated for thirty minutes reading the Romanian Communist Party newspaper. If Ivan did not approach me, I should leave. Another meeting might be arranged.

When meeting any Romanian, I took precautions. So, to keep my appointment with Ivan, I caught a trolleybus in the opposite direction to my destination, alighted after a few stops, took another in a different direction then walked to a tram stop. I let the first tram go, waited for the next, and rode it all the way to Herăstrău Park.

The sky was dark, the ground rutted with slush, so I picked my way carefully from to the park gates and thence to the restaurant. The pond was frozen and the air smelled of more snow to come.

Once seated, I began to feel the warmth. The restaurant smelled of hot plum brandy, beer and fermented cabbage. Two male customers were already seated at separate tables. Neither of them had been on bus or tram with me. I took a distant table and ordered what I always ordered in Herăstrău, "A beer and a plate of cheese and pickles." I opened Scînteia and crunched into the crisp dill pickle. From time to time, I glanced out at the snow, the bare trees, the empty paths.

Occasionally, hand-in-hand, a couple would stroll past. Then I spotted a small gentleman, shrugged into his overcoat, a wine-coloured scarf and a black karakul căciulă walk round the pond and take the path past the restaurant. He did this several times, apparently deep in thought. Finally, he opened the door, walked to my table and sat down. We shook hands as if we were old acquaintants filling in a dreary winter's day. No introductions.

I guessed that Ivan was some fifteen years my senior, my height, slim, dark hair brushed straight back. He had bright, penetrating eyes that seemed forever in motion and a wry smile as if he had already figured out precisely why what you were about to say was mistaken. He reminded me of an unblinking jackdaw, forever on the lookout for any curiosity that might amuse or be of use to it. Ivan, however, more engaging than any jackdaw.

Despite my asking him no questions, he told me quite a lot about himself. He was a Hungarian Jew from Timişoara. He'd studied philosophy first in Hungarian at Cluj and later, in Romanian at Bucharest. He'd joined the Communist Party, become a journalist, written novels and had a passion for puppet theatre.

One night in 1958 he'd been drinking with a group of Romanian friends at the Athénée Palace. A delegation of British dignitaries had entered. They were told that the bar was now closed. Most members of the delegation drifted off to their rooms, but one man sat down alone and without a drink. Ivan's friends left and Ivan invited the British man to share what was left of his bottle. They conversed for hours.

The visitor was a British MP. Following the Soviet suppression of the Hungarian Uprising in 1956, the Soviets had been under pressure to shed their reputation as political bullies. They withdrew Soviet occupying troops from Romania as proof of their benevolence. Romania had immediately grasped the opportunity to pursue a more independent foreign policy and had invited a delegation of British MPs to visit Bucharest to "encourage mutual understanding and trade". At the same time, to calm Russian fears, Romania adopted even tighter measures of internal security through its Secret Police and informant networks. Romania increased its punishments for crimes against the State including having any contact with Westerners.

Shortly after the British MP returned to the UK, he sent a postcard to Ivan with the words, "Thank you for your help." The Secret Police used the postcard as evidence that Ivan had acted in a way that threatened the security of the State. He was arrested, tried, found guilty, and sentenced to incarceration with hard labour. He spent six years in various prisons including Gherla. Gherla was notorious for holding political prisoners in underground cells at night and working them to death during the day. Ivan told me that he'd been freed just a few years earlier. Now he worked freelance for any Romanian newspaper that would hire him and wrote novels.

Ivan and I began to meet regularly. He was unique, interesting, and never asked me for anything. I enjoyed his company and was greatly flattered that such a worldly and well-educated man might enjoy mine.

Informants and the Writers' Union

My hiking trips didn't interest Ivan. My contacts with the Cultural Attachés in both the British and American Embassies did. Through the Writers' Union, he knew about the receptions that were held regularly for visiting British and American intellectuals. He envied fellow-Romanians who were able to attend these functions.

"Would you like an invitation, Ivan?" I asked, "I could suggest your name to our Cultural Attaché."

"I am invited to most of these functions," he told me, "But invitations come to the Writers' Union."

"So, why don't you attend?"

He took a mouthful of beer, "Invitations from embassies are sent to the Protocol Department of the Romanian Ministry of Foreign Affairs. They are delivered to the Communist 'base' at the Writers' Union. Trusted Party members select those who will receive their invitation or not. I am not selected."

"But, if you know you've been invited, why don't you just turn up?"

"If I were to attend without having received the invitation formally through the 'base', it would be used against me. The least they would do would be to expel me from the Writers' Union and then I would be unable to publish anything at all." The system was even more labyrinthine that I had imagined.

After such events as I attended, Ivan was avid to hear everything that the guest of honour had to say. He also wanted to know which of his colleagues from the Writers' Union had been present. They were those most favoured by the Communist Party.

After one such event I reported to Ivan that, "Only Zaharia Stancu was there." Ivan smiled his wry smile. Stancu, a poet, had been appointed President of the Writers' Union. As an afterthought I added, "Of course 'C' was there as well." 'C' had been imprisoned several times and several times expelled from the Union but had recently been reinstated. There was seldom a cultural function organized by Tony Mann to which 'C' did not turn up whether he had been invited or not.

"Why do you say 'of course'?" Ivan was interested.

"He turns up to every function. You know why." I paused without saying that 'C' was an informant.

Ivan gave another wry smile. "'C'? An informant? That's what you think?"

"It's what–" Abruptly I halted. Never did I share with Ivan or anybody else anything that I was told by the diplomats in the Embassy.

"So that's what your British diplomats think!" I kept silent. "I'll tell you what," Ivan was full of surprises, "next time you're at an Embassy function and 'C' is there, watch him. Don't leave before he does!"

The following week I attended a reception in Tony Mann's home for a visiting British academic. 'C' was there along with many other guests, some from the Writers' Union. I circulated widely at these parties to be as sociable as possible in return for Tony and Sheila's many kindnesses. Their apartment was exceptionally large. Guests wandered at will throughout spacious public rooms. Small table-lamps and the occasional standard-lamp provided discreet pools of light that encouraged small groups conversation. Waiters circulated with trays of drinks. Guests could help themselves to a cigarette from one or other of the elegant silver boxes on the coffee tables. There was an anteroom off the largest of these reception rooms, a space less popular except by those who wanted privacy to pursue a serious conversation.

As the party was drawing to a close after eleven o'clock, Sheila took my hands, "Darling, it's lovely to see you stay so late! You must be enjoying yourself!" I kissed her on both cheeks. To my credit, I'd quite overcome my embarrassment at the bussing and the squeals of 'Daawling'. Sheila and I talked pleasantly then she glided off to bid departing guests goodnight.

'C' must still be present, but I'd lost track of him, so I went in discreet search. He wasn't in the largest room and so I continued towards the far, dark anteroom. Nobody! Only two or three pools of subdued light and great expanses of darkness. I let my eyes adjust. Had I somehow missed 'C'? I was tired and walked to a corner where there was almost no light to think. Out of the corner of my eye I saw 'C's crooked form against the light of the larger reception room. He was alone. He peered into the darkness of the anteroom, and satisfied that it was empty, walked towards a coffee table. Tony himself had told me that 'C' was Securitate and so had the First Secretary. What I was seeing seemed to confirm this.

He crouched down and reached for the large silver cigarette box. Of course! The dead drop! Someone had left a message for him there; I was about to witness him retrieve it or perhaps leave a message himself. In a flash, all kinds of scenarios passed through my mind. Tony was an MI6 agent. He'd 'turned' 'C' and was now exploiting him as a source. My mind raced but I kept my eyes on 'C'.

After another furtive glance around, 'C' opened the cigarette box.

But instead of withdrawing an envelope or a slip of paper or depositing a roll of microfilm, he grabbed a huge handful of Tony's British cigarettes and stuffed them into his jacket pockets. I had to stop myself from laughing. Now all was clear. 'C' came to Embassy functions to satisfy his addiction to western tobacco! I remained motionless. 'C' patted his pockets to disguise the bulges and left the room. Moments later I heard his gentlemanly voice thanking Tony and Sheila: "A most stimulating evening!"

I left, slightly disappointed that Tony was not, after all, an MI6 agent using the cover of Cultural Attaché for secret purposes. Sheila deserved a good-natured British Council man and not a duplicitous spy!

A couple of days later Ivan and I met. "Did you go to the function?"

"I did."

"And 'C'? Was he there?"

"He was."

"You kept an eye on him?"

"I did."

"Do you still think he's Securitate?"

"He's a petty-thief!"

"Exactly! He attends to indulge his love for Scotch whisky and restock his supply of Western cigarettes."

"But how does 'C' dare to attend parties he's not even invited to?" I was puzzled.

Ivan gave me his knowing smile. "The Party has already done its worst to 'C'. He has nothing more to lose."

Ivan gave me food for thought. Surely the Party, the Securitate, the System could always find something more to take away, something whose withdrawal would serve to cause suffering or humiliation to body or spirit even to a man who'd suffered like 'C'. Were there those who truly had nothing more to lose? What about Ivan himself? He'd been a philosopher, a journalist, a creative writer; he'd been a political prisoner in Gherla. What more could *he* lose? I should have asked him.

The Editor

Ivan made one single request of me. When I told him that the editor of the Times Literary Supplement was coming to visit Bucharest he asked, "Can you arrange for me to meet him?"

"I'll guarantee you an invitation to the reception and to whatever presentation he's going to offer." I said.

"The invitation will not be passed on to me until it's too late!"

"Then what do you suggest?"

"Can you offer to take him somewhere interesting, like Herăstrău Park, early on the Sunday morning when nobody is about? Then I could meet you both at the restaurant there. Just for an hour!"

"Ivan, I can try to arrange that, but I will have to tell our Cultural Attaché exactly what I plan to do with him and who he will meet." If Tony agreed to ask Mr. Crook to accompany me on an excursion, both would have to know exactly what and who were going to be involved.

Ivan saw my point. I made my pitch to Tony. "I have a good friend, a novelist, his name is Ivan Deneș. He's a very well-read member of the Writer's Union and would dearly love to have an hour's conversation with Mr. Crook when he visits Bucharest."

"Could Ivan be Securitate or an informer?"

"I feel absolutely certain he's neither, Tony!"

Tony was sympathetic to my request. He asked me to draw up a precise plan. He would, he promised, present the proposal to the visitor as soon as he arrived for his approval.

Mr. Crook approved the plan. I called for him at the Athénée Palace. Together we observed my precautions and finally rode the empty tramcar to the terminus. It was just warm enough for us to choose a table outside. I'd noticed Ivan pass around the lake a couple of times. He would not join us until he was satisfied it was safe for him to do so.

Just as Mr. Crook wondered if my friend had been delayed, Ivan approached. We shook hands and I ordered a third beer for Ivan. Ivan immediately launched into a profound literary discussion. I enjoyed listening to their scholarly exchange on European literature.

The Kilt Behind the Curtain

After no more than twenty minutes, Ivan abruptly stood, shook hands and left.

"He's spotted a Securitate agent or a known informer." I explained. Crook made light of Ivan's sudden departure. I told him how I had refused to cooperate by refusing to police myself when I arrived and now, as a result, travelled as freely as I liked. He seemed to approve. I was, however, disappointed at how poorly the meeting had gone both for Ivan and for Mr. Crook.

The following day, Ivan asked if we could meet. "What went wrong?" I asked. "After all these preparations you must feel extremely disappointed. Was it Securitate?"

"No, I chose to end the meeting."

"You chose to end the meeting? Ivan, do you know what it cost me and the Cultural Attaché to set that meeting up for you?"

"Ron, that editor has read European literature only in English translation. I don't have time for anyone who hasn't read the Western canon in the original."

I was taken aback. Ivan excused himself – Romanians always had something to do – and left. As I walked home, I thought about his comment. It was true, many Romanians, all the educated ones I had met, were multilingual. Even mere bilingualism was considered a symptom of poor education. Ivan was an educated European who spoke four languages and could read in two or three more. He had suffered during the war; had studied at two good universities no doubt at great personal sacrifice; had spent six brutal years in prison. Despite being locked up, he had thought and discussed. Since being freed, he had penned pieces for major Romanian newspapers, written novels; found a role in the theatre, and read great literature. Ivan refused to understand how a person who had never experienced great privation could limit his reading to a single language.

I understood and so didn't hold Ivan's impatience against him. When Tony suggested that Ivan had been scared off by an agent of the Securitate, I nodded and said, "I expect so."

Years later when I was delivering a course to university teachers in Israel, I was reminded of Ivan's words about the importance of being multilingual. Each evening for two weeks, I ate dinner alone in the hotel where I was lodged. One evening, the maître d'hôtel met me at the door with an apology.

"Professor Mackay, I'm sorry but there's a reception here this evening. I must ask you to share a table."

"Fine," I said, "I'll be glad to have company for a change."

"Thank you. Tell me what languages you speak so I can find you compatible dinner companions."

Multilingualism, not mere bilingualism, was an automatic assumption!

A Day in the Life of Belu Zilber

Ivan invited me to his home. He was recently remarried to an ethnic Romanian younger than himself. They and their baby had been allotted an apartment in a new development on the outskirts of the city. I walked in mud through a building site to one of three mammoth apartment buildings.

"Ivan is in his study."

The room was a study-cum-bedroom. Beds without covers, served as settees. Never having been able to sit comfortably on a bed, I chose an upright chair. I'd no sooner sat down than the doorbell rang again and Ivan's wife admitted another guest. *Should find an excuse to leave?* The door opened and in walked a fit, watchful man in his late 60s. By the way Ivan and he greeted each other it was clear they were close friends.

The gentleman turned to me and extended his hand, "Belu." I took it, "Ron."

Belu immediately made himself comfortable on the divan and began asking me questions about British politics. I was unable to answer competently. Belu assumed that my handicap was Romanian and so he kindly switched to English. He soon realised that my limitations were both linguistic and intellectual. I simply hadn't a sound grasp of political philosophy. So, Ivan and Belu talked. They switched effortlessly between Romanian, Hungarian, German and French. For my benefit they summarised in English.

Seeing Belu and Ivan sit there in total comfort, calmly discussing the political issues of Western Europe and Eastern Europe, two very different points occurred to me. The first was that both men must have spent a long time in prison to sit so comfortably in that semi-reclining position. The second was that here in this small, utilitarian apartment on the muddy outskirts of Bucharest, these two men were holding an informed, profound conversation that would have been a credit to a salon in London, Paris or Berlin. They had been and may still be communists, I thought, with nothing of material value to their names, but they could take their place as peers with any group of intellectuals in any western capital. The intellectual resources of so many I met in Romania awed me.

After Ivan's wife brought us coffee, I left. Belu did not leave with me. It was one thing meeting me – most certainly by design – but travelling with me on public transport or being seen with me, was out.

A few days later, 'M' said that Ivan had spoken to 'D' about my visit and the conversation with Belu. 'M' was clearly impressed that I had met Belu and told me that both he and Ivan had been close friends of her father's and her mother's before, during and after the war.

"Was Belu in prison with Ivan?" I asked.

'M' told me that Belu was a Romanian Jew from Iași. He had been in political trouble all his life mostly because of his communism. He'd been imprisoned under King Carol II for spying for the Soviets; imprisoned again after the War; expelled from the Communist party for criticizing Stalin. Together, he and Ivan had survived the notorious Gherla prison.

"Belu was recently pardoned by Nicolae Ceaușescu. Now he's fighting to regain his membership in the Communist Party." 'M' merely smiled at my astonishment. Romanian minds were used to balancing mutually exclusive ideas. In my Scotland, things were either black or white. Not so in Ceaușescu's Romania.

Who is Ivan Deneş?

After completing my two years in Bucharest, I heard that Ivan had left Romania around 1970 and settled in West Germany. At the time, I held a lectureship at the University of Newcastle upon Tyne. I planned to visit Ivan in Berlin and also Harald Mesch who had defected to West Germany in 1969. For some reason, the visit never happened. I lost touch with both Ivan and Harald.

However, my research tells me that the man I knew as Ivan Deneş was born Iván Alexandru Deneş, a Jew, in Hungarian Timişoara. He became an informer for the Romanian Secret Police in 1948 and worked as an informer and spy for the rest of his life. In Romania he used various aliases 'Aurel Bantaş', 'Alecu Sirbu', 'Alexander Sirbu'. He was sent by Romania to Israel as a spy with the code-name "GX-36". He adopted the names Peter Pintilie, Kraus and Konrad in Germany where he worked for Axel Springer.

Ivan infiltrated Radio Free Europe but was exposed as a spy by employees who had shared prison cells with him in Romania. Ivan was fired. He returned to work for Axel Springer, somehow managed to continue to make broadcasts for Radio Free Europe under the pseudonym Ion Daniel, and founded his own news agency, Ost-West-Presseagentur.

Ivan visited Romania often and, in 1989, nominated for a prestigious award by Aristotel Stamatoiu, Romania's Interior Minister and Director of the Central Foreign Intelligence Directorate of State Security.

Thrice married and divorced, Ivan died an alcoholic in Berlin in 2011. He was 83.

There is a telling quote from 18th century Romanian nationalist Tudor Vladimirescu, that bears some thinking about in the case of Ivan Deneş: *"There is no law that would prevent a man from meeting evil with evil."* Just as that sentence allows for multiple simultaneous interpretations, so does the character and life of Ivan. Ivan was a man in whose mental make-up and daily life there were such unpredictable sequences of changes that it has been impossible for me to fathom who or what he really was.

I remember Ivan as the most interesting man I have ever met and whose company I have most enjoyed. I believe we were friends who liked and respected each other.

Ivan as he appeared in his security file.

42

SITTING DUCKS

From time to time Sică Stevoiu would call me at my apartment. I'd met him and his wife Anica in their private villa in Sinaia.

Sică would ask what I was up to. I always found something interesting to tell him. My rule was to talk about my travels, never my relationships with others. "Why don't we meet up for a coffee?" was his invariable suggestion. We would arrange a time and a place but when the day came, only his wife Anica would turn up. Anica and I would sit together at a small table in the pastry-shop that smelled of Turkish coffee and talk. In Anica's company, I never gave Sică a second thought.

The immaculately groomed Anica was older than me and aroused a confusion of feelings. I was enraptured. She was good-looking, dark-haired, and her bearing spoke of unlimited confidence but no arrogance. Anica looked directly into my eyes when I spoke. She would touch my arm when making a point. A strand of hair would fall over her face. I wanted to gently brush it back, but I never dared.

We talked about everything without the conversation ever turning personal or political. I'd worked my way through university at a variety of unusual jobs and had lived in several countries and so had unique tales to tell. She would listen intently as if there were no one she'd rather be with. When my Romanian ran into difficulties, she would provide just the right word and nod for me to continue. We both knew the boundaries and respected them.

I felt hugely flattered to be with such a woman. No coffee-drinker I, but in Anica's company I could match her coffee for coffee the entire afternoon. I was enamoured.

"Sică has invited you to a party in our home." Anica explained that, after the party, at dawn the following morning, Sică would take me and other male guests on a duck-shoot. I should come dressed warmly and prepared to stay over at their house.

I cared little for parties. Out of a feeling of duty, I attend a few at the British and American Embassies. The idea of going to a party where only Romanian would be spoken and only Romanians in attendance, and from which I couldn't escape because of an all-male hunting excursion did not appeal. As I was about to refuse, Anica put her hand on my arm. I felt her warmth. My eyes were drawn to a moustache of icing sugar on her upper lip.

A doting eye exercises little judgement. I heard myself saying, "I would love to accept, Anica!"

I'd handled most infantry weapons, but I'd never fired a shotgun. What bothered me more, however, was the idea of careless men handling loaded guns after an alcohol-fuelled party.

"I'm not very good at late nights." I was seeking a way out.

"Don't worry," Anica reassured me. "You can sleep if you want before you go off with Sică to Snagov." I'd heard that Ceauşescu and his Central Committee had a private estate on Lake Snagov.

Before we parted, Anica gave me their address. "Come in Romanian clothes. Take a taxi to within two blocks of our house." She gave me the name of the intersection. "Walk from there. Bring your University ID." This was Romania, bizarre instructions were commonplace.

The evening of the party arrived. I gave the driver the street intersection. The driver looked at me. "You're sure this is where you want to go?" I confirmed. He shrugged. We drove to a part of Bucharest I had never visited, a hint of private houses behind high walls. There was an unusually high number of uniformed Milițieni. From the intersection I began to walk the two blocks to Sică and Anica's home.

"Comrade, your I.D. please!" He spoke into a two-way radio and escorted me to a solid wooden gate in a high wall. Two plain-clothes Securitate offers confirmed my I.D. and allowed to enter. I found myself in an attractive courtyard with gardens beyond. The door of the main house stood open in welcome. Two plain-clothes officers checked my I.D. yet again and gestured me into a brightly lit reception room. Photographs lined the panelled walls. I could see lots of people, all Sică and Anica's age, some older. Sică detached himself from a group. He made his welcome sound as if my coming were a favour. Briefly, he talked to me alone then introduced me to a group. "These men will be hunting with us at dawn."

I was out of my depth. Despite having been in Romania for eighteen months, my Romanian was poor. My professional contacts at University insisted on speaking only English. I could function adequately in everyday situations with shopkeepers, peasants at the market and the Miliție, I could request directions, buy train tickets, order meals, and register myself into a hotel. I could hold limited conversations with fellow hikers up in the Carpathian Mountains. Here, however, among highly privileged Party members, I felt at a distinct disadvantage. To my surprise, none appeared concerned that I was foreign.

Doing my best to look at ease, I wandered round the room examining the photographs on the walls. Many showed the same man. Anica appeared and tilted her cheek for a kiss.

"You're all alone and you haven't got anything to drink!" She led me to a table, poured us each a drink and drew me into a quiet space. Sică joined us.

"I was examining the photographs." I said.

Surprise, Surprise!

"The photos are of Anica's father." Together, they escorted me round explaining each frame. Many were from immediately after the War when sympathisers helped bring the Communist Party to power.

"Here is Anica's father leading a protest at the railway workshops. Here he is encouraging striking workers. Here, he's showing the Party the way forward."

Then, her father was a well-dressed young man who carried a walking stick as if it were a sword. Close to the entrance hall, Sică asked the two Securitate officers to move aside. There, in a glass case, was the walking stick we had just seen in Anica's father's hand. It was a handsome, wooden cane with an elaborate handgrip.

Sică opened the case, removed the cane, and while holding the handle, offered me the stock. As I gripped the stock, Sică tugged. I was left with a hollow wooden case but now Sică held a metre-long, steel blade.

"A sword-stick!" Sică explained. "He was always at risk. He needed this to defend himself."

Sică drifted off. Anica continued to tell me about her father. Like many Romanians after the War he had embraced communism and soon became a leader. He knew Petru Groza, the first post-War Prime Minister of Romania who helped the Communist Party to seize a grip on Romania. He had held several high positions. When he had died, he was a director of one of the branches of the Department for State Security of the People's Republic of Romania. His memory, Anica told me, was revered.

"This was his and my mother's home. Now it's mine and Sică's." I guessed there was more to the story, but it would have been imprudent to ask.

Exhausted by all that I'd done and heard since arriving, I asked Anica if I could rest. She had a Securitate officer escort me to a suite of my own.

"You will be wakened an hour before dawn," he told me. I fell asleep in my hunting clothes.

I was duly roused. We clambered into a convoy of shiny, black Russian limousines to head to a lodge in an estate on the shores of Snagov Lake. Unlike some of the hunters, I'd slept well and drunk little. Sică and I settled into the back seat of a large black Chaika. He looked the worse for wear but unconcerned.

Not a Word!

"Don't say a single word when we get to the lodge," Sică warned me.

A score of tired men struggled out of the limousines and entered the magnificent timber-built hunting lodge. Heads of antlered deer and tusked boar decorated the walls. Several black bears stood with outstretched arms and showing ferocious teeth. Liveried waiters served us breakfast. Finally, the waiters brought plum brandy in tiny glasses. I sipped mine to help to stave off the cold.

"You are with me, Ron. I'll put you in the bow, it's the best position to fire from. There will be two more guns amidships. I'm in the stern."

My military training had given me a healthy respect for all weapons and especially loaded firearms. I had no idea how experienced Sică or any of the others might be. They appeared to have all the right clothes, but the amateur can confuse appropriate clothing with technical proficiency. I wasn't willing to take any chances with individuals unskilled in respecting arcs of fire.

"I'm sure you're a better shot than I am, Sică. Why don't you take the bow?" I believed that I was less likely to blow somebody's head off than was he or one of his chums.

Uniformed gamekeepers presented the guns correctly, stock-first with barrels broken open so that the recipient could see the empty chambers. I was handed a beautiful, Czechoslovak double-barrelled shotgun. The ghillie who would be in our boat went over it with me. It was almost identical to the guns I had cleaned, but never fired, in Scotland.

The ghillie in each boat would carry the cartridges and would hand these out only when the boats were in position in the distant reeds. He

The Kilt Behind the Curtain

showed us how to carry the guns to the boats. I saw our ghillie remind Sică, to return the gun to the cradle position after he casually slung it on his shoulder. The gamekeepers were expert. Nevertheless, I was glad that I had decided to take the stern position.

Lots were drawn for positions around the shore of the lake, then each group boarded its boat and we quietly motored off to our allotted positions. It was still pitch black. We cut the engine and floated into reeds. The only sounds came from waves lapping the side of the boat, roosters announcing dawn, and the occasional far-off bark from a farm dog. I have always enjoyed daybreak, watching the sky gradually lighten in the east, hearing the countryside wake up. Sică and the other two guests crouched below the gunwales out of the wind and tried to make up for their late night.

As the eastern horizon began to turn pale yellow, we could hear ducks whirring in to land but none near us. Nevertheless, we loaded our guns. The ghillie reminded us of our arcs of fire. An explosion of distant shots came from several different points on the lake. Still no ducks close to us. Sică and his guests were becoming irritable. Suddenly we heard wing beats overhead. All the guns in the boat, except mine, went off at once. They hit nothing but all were excited by their "near misses".

This routine was repeated several times with the same results. It was now almost fully light. We had no more than ten minutes of opportunity left. Sică insisted on changing places with me. He desperately wanted me to get a bird, so I moved to the bow.

Suddenly a pair of birds appeared out of the reeds twenty yards ahead of us. I raised my gun and slipped the safety off. When I saw that they were great-crested grebes, put the safety back on.

"Get them, Ron!" hissed Sică. When I paid no attention and he saw the birds were moving away from us out of range. "Fire!"

"Rațe nu sunt!" I whispered. "These are not ducks."

"I don't care! Shoot them!"

I refused. Sică sulked all the way back to the landing. We were the only boat with absolutely nothing to show for our dawn hunt.

I'm sure I saw the ghillies exchange the same looks that Scottish gamekeepers swapped as their guests left to be driven to the train station to catch the overnight sleeper back to London.

"Thank the Lord we're done with these damn toffs for another year! Now our lives can return to normal!"

After that episode, I enjoyed coffee with Anica only two more times. I relished her hand on my arm, the wisp of hair over her face, her white sugar moustache and the flakes of pastry on her lips.

"I'm leaving very soon!" I told her.

"We are going to miss you!" Those amber eyes.

I chose to imagine she used the royal "we".

Pearl at Lake Snagov in 1969. The communist dictator Nocolae Ceauşescu had a hunting lodge closeby.

43

TIBERIU STOIAN AND THE SCHOOL FOR SPIES

Accosted!

The simplest way for any Romanian who wanted to contact me was to lie in wait for me as I left the University. I was a creature of habit, invariably returning to my apartment after class. I'd given up using the University Faculty Club on Strada Dionisie Lupu. I'd find myself quite alone at a table for four. Colleagues could enter the dining room, look around and leave. My table was simply "out of bounds" to them. I found this as embarrassing as they did and so did all of us a favour by simply going home.

I invariably turned right onto Pitar Moş, right again onto Strada Rosetti and then left onto Magheru Boulevard. If I were walking home, I stayed on the same side of the road all the way back to my apartment building over a mile away. If I decided to take public transport, I had to cross the main boulevard to a bus-stop. Anybody wanting to contact me needed only to know my teaching schedule and then they could find me with ease and follow me.

That's exactly what Tibi did.

As soon as he turned to address me, I knew I'd seen him twice before. He was about my age, taller than me and had a shock of uncommonly blond hair. I thought he might be Securitate.

"Professor Ronald Mackay?" He had a friendly, confident smile.

"Yes!"

"Tiberius Stoian! My friends call me Tibi!" His English was perfect. I shook his offered hand.

I'd learned to assess Romanians very swiftly and decided that he was more likely to be an informer than an agent of the Secret Police.

"Is this an inconvenient time?"

"It's lunch time, I'm going to have lunch, would you like to join me?" He could accept it or reject my invitation. Informer or not, his company was better than eating alone.

"Thank you!"

Together, we walked to the quiet restaurant where I occasionally lunched, invariably alone. I welcomed his company. He was forthcoming. He told me he'd been a student in the very faculty I taught in and, after graduating, had worked for "Radio Romania" but now worked as an English teacher. I filed that information away as extremely odd. Posts in State radio were offered only to the best graduates, school teaching to the less gifted. One of his parents was from Transylvania and the other from Bucharest. That helped me explain his looks.

He asked me if I knew of any opportunities for internships with the BBC in London.

"I could find out for you," I offered. "It should take no more than a week."

"If you agree, I will meet you next week, in this restaurant at the same time."

There were internships with the BBC in London. Tibi shrugged when I told him. He elaborated his background and his family. His father was Romanian, his mother Hungarian. I found him good company and when, after lunch, he asked for my telephone number I gave it to him.

If he felt comfortable calling me on a line that was bugged, that was up to him.

The following week he called and suggested we meet in Cişmigiu Gardens. It was a beautiful late spring day. Flowers were in bloom and the trees freshly green. We walked until he found an isolated bench.

"I am going to defect. In London. I will work for the BBC."

I was taken aback at Tibi's candour. I said nothing but wondered if I was being set up.

At that moment, a huge bumblebee floated towards us and hovered over a fragrant lavender. Tibi looked at me with wide eyes, put his finger to his lips and pointed to the bee.

"Secret Police! They're everywhere!" We burst out laughing. His sense of humour dissolved the tension.

Tibi told me that he'd excelled in his English studies, had loved his work with Radio Romania but that the Securitate had 'borrowed' him from State radio and installed him in a school where they trained agents to undertake missions abroad in countries where fluent English is essential.

If I'd been taken aback by his earlier candour about defection, I was dumbfounded by his admission that he trained spies for the Romanian Government. I let him talk.

His intention to defect was firm, despite the excellent salary and privileges he enjoyed.

"Will you give me your address in the UK?" He already knew I'd soon be leaving Romania. I gave him my mother's address in London, but told him that I would be in Edinburgh studying for a postgraduate degree. I made it clear that I would be unable to help him in any material way.

"I don't need financial help," he assured me, "I just want somebody I know in the UK when I defect." Having experienced loneliness, I understood.

I heard nothing from or about Tibi from that day in May 1969 until two-and-a-half years later. But that's quite another story.

44

WHAT PLANS, RONALD?

Pulling Up Stakes

"What plans do you have for 'M' after you leave Romania, Ronald?" Pearl was paying a final visit to Bucharest. Soon, I'd return to the UK to study for a postgraduate degree, almost as penniless as I'd left.

While both 'M' and I knew that we would soon part, we hadn't formulated any explicit plan. Our paths had happily crossed but neither of us honestly believed that our coming together was other than temporary.

'M's and my worlds might share a little in common by virtue of having coincided behind the Iron Curtain at a critical time. We each knew our relationship to be a joy in the palpable present. Neither saw it as enduring into an uncertain future.

As a very bright, healthy, good-natured, and ambitious nineteen-year-old, whose reality was Post-War communist Romania, 'M' focused and thrived on the present. She was her own heroine in her own drama. I happened to make a cameo appearance. There had been and undoubtedly there would be other characters to play supporting parts to her lead role in the years to come. 'M' had never said as much but her anecdotes, observations and behaviour, led me to this understanding.

Each of 'M's encounters was an honest and intense adventure. With her high-school sweetheart; with the notorious Lothario in her group of teenage friends; with classical musicians she was occasionally contracted to interpret for on behalf of State Radio and Television; with me.

Women students in the University often cast me flirtatious glances. 'M' had commented on this many times, not, as I had first imagined, out of jealousy but out of personal satisfaction. So many women would have welcomed my attention, at least the attention of the "Visiting Professor" - and yet it was she, 'M', who prevailed. That knowledge gave her a heightened sense of her power as a woman. It gratified and validated her amour-propre. She eclipsed her peers and outcompeted her rivals in every possible way.

This truth was driven home to me one day when 'M' asked, "What do you think of Romanian girls?"

"I think they're exceptionally beautiful and very romantic. They flirt daringly."

"Do you flirt back?"

"No." My life was made simpler if I enjoyed the looks cast at me without encouraging escalation.

"I'm your only girl-friend?"

"Yes." I was troubled by the question. Might 'M' imagine I was a philanderer?

And then she hit me with a heartfelt pronouncement that dumfounded me, "With all the opportunities you have, you should be cheating on me all the time!" Shocked and embarrassed, I laughed but for days after, I pondered that remark. What did it say about her and about me? We were worlds apart.

The conclusion I arrived at was that my loyal behaviour lacked the romantic drama and the competitive challenge that 'M' sought. As a dynamic and vigorous woman, might she have preferred the cut and thrust of hand-to-hand combat to the respect I offered her. It was then I saw that 'M' played the champion in her own personal narrative. Those who played the supporting roles could be important but were interchangeable if they failed to live up to the exacting level of performance she expected. If the action changed and they no longer suited, they could be replaced. 'M' was at that stage in her young and intellectually privileged life where she was trying out new roles for herself. She possessed, not without justification, the supreme confidence that she could play any part with outstanding success, aware – and rightly so -- that she possessed impressive talents.

I groped for the words to explain this to Pearl. She waited until my efforts stumbled to a halt.

"No plans? That's not what 'D' thinks." She looked at me seriously. "I have the distinct impression that 'D' would like to see you and 'M' marry."

Now that Pearl had made that explicit, I was faced with something that I'd been unconsciously sidestepping for some time: 'D's agenda. 'M' might live in the present and play the starring role in her daily narrative. 'D', however, was playing a different role, that of protective mother. 'D' had descended into present-day life in the People's Republic of Romania from a more elevated and vastly different past. She was surviving thanks to her skills and her acumen. Like thousands of Romanians who had known better days, 'D' nursed ambitions for her daughter and had every intention of seeing these ambitions fulfilled whatever the sacrifice, whatever the cost. At least her daughter – and perhaps in the long-run she herself – might return to a more promising, less punishing world. Who could blame her?

Pearl, likewise, was protecting the interests of *her* off-spring. She would not stand by and see one of her brood used as an unwitting pawn in a private game of chess. Pearl needn't have worried. It had become clear to me that I must be sufficiently explicit with 'M' that the end of my contract also meant the end of our relationship. Neither of us should court the risk of joining the ranks of the walking-wounded.

I was confident that I could make explicit to 'M' what had been, to date, an unstated understanding without inspiring any overly dramatic reaction. I was right. 'M' smiled her glowing smile as if to say, "Why must we unnecessarily repeat ourselves?" and took the opportunity to ask me if I would give her my second-hand, portable typewriter as a parting gift.

My portable typewriter was the only possession I had of any value. It might have been worth, perhaps £20. 'M' carried it home with her disguised in a shopping bag. If typewriters were not banned outright in this Communist country, they were not something you wanted to draw others' attention to. Now I had nothing more than my 35 mm camera, the clothes I stood up in and those few hanging in my wardrobe.

Two days later, she brought the typewriter back. "Can you have some of the keys modified to provide Romanian diacritics?"

If I needed any confirmation that I had made the right decision, this was it. 'M' and 'D' were two to my one, and so it always would be.

I duly gave the typewriter to Ron Walker, the Hymac engineer, the next time he was driving back to the UK via Vienna. He arranged the exchange of keys that 'M', or more likely 'D', had requested, brought the machine back to Bucharest, and I duly delivered it to 'M'. The modifications cost me 15% of my annual Sterling income. The only possible use my typewriter could have for 'D' would be as an item to barter in exchange for some service of extraordinary value. Perhaps an exit visa?

Comparative Living Standards

Most if not all Romanians assumed that everybody in the West was well-off. The truth was that we were better off in terms of political freedoms but we, in the Scotland I knew, were not significantly better off in certain material terms. It was impossible to explain this convincingly to a Romanian. He or she believed that you were merely trying to save them the embarrassment of acknowledging their own privation.

As a summer teacher in Bournemouth, my standard of living would take a drop from what I enjoyed in Bucharest. I would return to a single bed-sitting-room in a lodging-house where the owners resided in the basement and rented out the other five bedrooms individually. We would all share the kitchen and bathroom. I would be able to go out for entertainment once a week, to the cinema or a pub, to a dance, or for an Indian meal. I would wash my own clothes, press my trousers and iron my shirts; do my own shopping; prepare a sandwich for lunch, and make my own dinner. As a lover of the spoken word, I would allow myself the luxury of a small transistor radio, but no TV. I would use the communal payphone; no long-distance calls; no luxuries and no vacation. That was what my monthly £200 would permit me if I wanted to save for a year's post-graduate study.

One more critical conversation with 'M' took place before I left, in my apartment. With the bathwater running to confound the surveillance microphones, she told me that she was being excluded from even applying for the British Council scholarship unless she "cooperated". The secretary at her "base" found a single lowish grade in geography from an earlier year. However, they'd make an exception if 'M' showed her loyalty to the country by joining Communist Youth.

'M' knew that this was a foreshadow of her future in Romania. Opportunities would be traded for proofs of loyalty. Who knew to what degree? Not a pleasant prospect.

Acknowledging this danger 'D' promised to mobilize every connection she had, to obtain exit papers for her daughter. For my part, I offered to write a letter of recommendation to Edinburgh University, attesting to 'M's outstanding academic performance and her political circumstances.

"Using my title as British Visiting Professor, Edinburgh University, may allow you to enroll."

I considered the chances of 'M' obtaining an exit visa to be slim, but why not at least open a door for her in Edinburgh? If anyone could make it, it would be 'M'.

45

FORMAL GOODBYES

Tony and Sheila Mann offered to host a cocktail party in their home before I left to say 'Goodbye' principally to my University colleagues and Embassy personnel.

"I appreciate the gesture, Tony," I told him, "but I've exchanged no more than a 'Good Morning!' with most of my colleagues. I can say goodbye to my friends myself."

"Protocol demands it, Ron!" So, Tony and Sheila organised the party to celebrate a milestone in the British-Romanian Bilateral Cultural Agreement.

It turned out to be a formal 'Who's Who?' of the faculty powerbrokers. Madame Cartianu and Professor Chițoran appeared. They were entitled. Others I had particularly asked to be invited, failed to appear. The more confident Professors Levițki and Duțescu were there as senior scholars who contributed greatly to the University's reputation. In Edinburgh the following year, I was able to host them and their wives for dinner. They were on a British Council-sponsored trip to the UK.

My friend Dr Dino Sandulescu was present, cutting through inane small talk to make a significant intellectual point. Harald Mesch was there, quiet, confident, alert. I was happy to see Dean Ion Preda, he who had introduced me to the fiery chili peppers! Happily, the gentle Professor Ştefanescu Draganeşti and his wife attended. To my surprise, Professor Tatiana Slama-Cazacu was there with her silent husband Boris. In two years, I'd encountered them twice and only briefly.

Tony and Sheila and their young daughter Sarah who had just finished the semester at her boarding school in England, judged the cocktail party marking my imminent departure a great success. I was glad. Tony had been my rock. I greatly appreciated and benefited from his and Sheila's friendship.

After my last class, I exchanged warm handshakes with all the junior professors who had come to say goodbye. Then, without fanfare I walked out onto Pitar Moş for the last time.

Those few friends I had made had each cautioned me. "Do not write!" A letter from the West could lead to a visit from the Secret Police followed by unfortunate repercussions.

TAROM's BAC One-Elevens

I had a ticket to London on one of the British BAC One-Elevens that TAROM, the Romanian state airline, had purchased from the UK. A few weeks earlier, I'd had dinner in Doris Cole's apartment with the two British pilots who were teaching their Romanian counterparts to fly the new aircraft. They amused the guests by acting out the challenges that these Romanian pilots, used to flying the slow Soviet Ilyushin turbo-prop aircraft, encountered with the much faster BAC jets.

Take Off

My plane rose high over Bucharest. I looked down at the city spread below; the city where I'd spent the better part of two important years of my life. The aircraft adopted an easterly heading towards Hungary. The plain to the south stretched towards the Danube and the border with Bulgaria; to the north were the forested Carpathians where I'd spent so many enjoyable weekends in all seasons and in all weathers. I was leaving friends, a lover; probably forever. I was leaving a way of life that I was unlikely to experience ever again. I reflected on the wonderful experiences, on the frightening and on the shocking ones. I thought on the good things I'd seen and on the bad. It had not and still was not easy to assimilate the diversity I'd lived; no simple matter to winnow the wholesome from the pernicious.

46

SUMMER OF '69

Bournemouth

I began teaching for the second summer at the Anglo Continental School of English in Bournemouth, living frugally to save for my year of postgraduate studies at Edinburgh University. I was determined to accomplish my goal and knew that somehow, I would.

'M's Triumphant Arrival

One evening in early September the lodging house payphone rang.

"Ron – it's for you!" I took the handset from him. He mouthed. "A woman! Lovely voice!"

"I'm in London!" 'M' was ecstatic. I'd given her the boarding-house number before I left so she could contact me if she were fortunate enough to be awarded one of the scholarships or to obtain a visa. "I received my exit papers! I'm here!"

"Congratulations!" I was delighted for 'M'. She was a winner.

"I want to come to see you this weekend."

"As a friend?"

"As we agreed. Friends."

When I met 'M' at Bournemouth railway station, she was literally bubbling with joy. We dropped her tiny bag off in my room. She had the single bed; I a blanket on the floor.

We explored Poole Harbour.

"Your letter worked! Edinburgh University have given me credit for one of my two academic years." She would complete a first degree in English literature with German as her 'second' language. I smiled at the British commonplace, after your *first* language you might master a *second*. 'M' mastered five!

I too had opted for Edinburgh University, to study in the Applied Linguistics department. My programme advertised its mission as to train graduate students in the resolution of educational problems rooted directly or indirectly in an inadequate command of language. As a result of my experience in Romania, I was beginning to see myself first and foremost as a *'problem solver'* – someone with the resources to address the questions of others, to examine and assess their concerns, and to offer practical, principled ways to resolve them.

Here We Are Again, Alexandru!

Over dinner, 'M' told me about the challenges she faced in preparing to leave Romania. 'D' had set in motion everything she knew about how to obtain 'M' a passport with a tourist visa good for two weeks in the UK to perfect her language skills. Leaving a family member behind as security was a favourable element in such applications. To obtain the coveted passport and exit visa, 'M' had to be interviewed by the Secret Police at Securitate headquarters. She had waited nervously in an anteroom. Finally, she was called into the official's office. There, seated behind an impressive desk over which presided the omniscient portrait of Ceaușescu with his weak chin and crooked grimace, sat none other than Alexandru from Alexandria!

Alexander was dressed resplendently in the uniform of a Major. Among many other questions flung at her, 'M' was asked:

"What do you know of Professor Mackay's counter-revolutionary mission in Romania as a British undercover operative?"

She knew nothing, of course. What was there to know?

'M' passed the interview. 'D' put her on the train to Vienna within 24 hours of receiving the exit visa. It was not uncommon for those to be rescinded at a whim or based on a neighbour's jealousy. When the train left for Vienna, neither knew if or when they would see each other again. I cannot imagine a greater sacrifice a widowed mother can make than putting her only child on a train, never to return, so she might seek better opportunities. 'D' would be interrogated for months by the *Securitate* about the "terrible education" she had given her daughter, now a defector. It would be years before 'M' and 'D' would be united.

Having been an interpreter for visitors from the Vienna Opera and the Vienna Burgtheater, 'M' reached out to some of the artists on this first stop. Touchingly, one of the actresses put her up for several days and collected a few hundred shillings for her.

In London, her cousin introduced her to a Jewish philanthropic group who offered immediate practical assistance to refugees from Eastern Europe. They helped in several ways – immediate accommodation and food, clothes, and a little money. 'M' had left with a suitcase that carried what had to look like just a summer's wardrobe. 'M' was already adapting successfully to her new environment. Later, her American relatives, who had emigrated to the US in the '30s and were well established, helped her out as well.

'M' and I were on different paths through life. These paths had temporarily coincided, and for both of us happily and unforgettably, in Bucharest.

47

ROMANIAN FALLOUT

'M's Destiny

The last communication I had from 'M' was a letter she wrote to me in 1975 when I was running a British-funded project in Mexico. She had migrated to the California that she had talked to me about since I first met her.

'D' had already settled in California, having migrated to Israel from Romania in 1971, where she met and married an American doctor. Unable to learn Hebrew well enough to practice their professions, they settled in the US. 'D' cultivated a highly successful professional practice offering master classes to established musicians. 'D' was indomitable.

Showing her own entrepreneurial capacity and intellectual curiosity, 'M' switched professions, left academia for the "real" world of banking. She held several executive positions in California. Today she is the president and co-founder of a dynamic consulting firm that delivers leadership development to global business leaders. She is also a director on the board of a publicly held international corporation.

I admire 'M' unconditionally. I applaud her for all she ever was, has grown into, and has accomplished in this challenging world – her love of life and her talent, her ambition and vision and her tireless application. A truly remarkable woman, 'M' will always command my affection and my esteem. Reconnecting remotely later in life, the foundational experiences of those early years remain precious to us both.

Suspicion and Secrecy

The first unexpected fallout from my two years behind the Iron Curtain was in 1978. At that time, I held a teaching appointment at Concordia University in Montreal and was living on my nearby farm in Ontario. One afternoon I drove home from the railway station to find a pickup-truck parked by the farmhouse.

Two men were examining the interior of an old drive-in shed where I kept my tractor. On seeing me arrive, one of them walked, unhurried, to my car to engage me while the other continued his search for something high up among the rafters.

"We're interested in buying the farm."

"The farm's not for sale."

"Oh? We were told it was." I knew he was lying. His partner, search completed, joined us. Both were kitted out as all farmers were in Glengarry, in matching green workpants, green shirts and unlaced Kodiak boots. Their clothes and their boots, I noticed, were spotlessly new.

"Army or police," I reckoned. "What could they want with me or my farm?"

They turned out to be officers of the RCMP, the Security Service branch that undertook counterintelligence operations in Canada. I was suspect. All this, they told me much later.

For several years, they called me or simply appeared at irregular intervals and quizzed me about my political views, my friends in Canada, Mexico and in Romania and my current activities. At their bidding, I took all my Romanian photographs and transparencies to their headquarters in Ottawa. They scrutinised each, one by one.

But that is a whole other story!

ACKNOWLEDGMENTS

For their contributions, comments, and plain enthusiasm, my thanks to: Viviana Galleno; Vivian and John East of Wilmslow, Cheshire; Dr. Euan Lindsay-Smith of Brisbane, Queensland; Dr. Palmer Acheson and his wife Lise, of Lethbridge, Alberta; Peter Elmhirst and Anne Marshall of Elmhirst's Resort, Keene, Ontario; Mariana Marinescu of Bucharest and now the US; Loredana Crina Iacome (Crina Soimu) of Peterborough, Ontario; and Skee and Geza Teleki, of Cramahe Township, Ontario.

Diane Taylor, author of "The Gift of Memoir", whose workshop in 2015 inspired me and whose advice I followed.

I owe special gratitude to: Dr. Doina Lecca of Montreal, Quebec, who lived in Bucharest and studied at the Faculty of Foreign Languages and Literatures during the time I write about. While not always in agreement with me, Doina recognized my account of ethos, characters, and events as I wrote them, chapter by chapter.

Also, to Dr. C. George Sandulescu and Dr. Lidia Vianu for publishing the original, longer manuscript of this book in 2016 in their

Contemporary Literature Press, the online Publishing House of the University of Bucharest.

Finally, thanks to Victoria Twead and her team at Ant Press for converting my manuscript into this lovely book with such skill, speed and courtesy.

ABOUT THE AUTHOR

The author and his brother, 1954.

Ronald Mackay was born in Scotland during World War II. After leaving Romania in 1969, Ronald pursued "problem-solving" as a career. With a doctorate in programme evaluation, Ronald Mackay held posts in several Canadian universities and worked internationally to improve the planning, management and evaluation of development projects in agriculture and in human well-being. He has farmed in Canada, Mexico, Chile and Argentina. In 2012 he turned to writing memoir, drama and short stories. He is the author of "Fortunate Isle, a Memoir of Tenerife" and "A Tenerife con Cariño". Ronald and his wife Viviana, live in the house they built on the shores of Rice Lake near Keene, Ontario, Canada.

Printed in Great Britain
by Amazon